Lu Xun and Evolution

∞

James Reeve Pusey

State University of New York Press

Published by
State University of New York Press, Albany

© 1998 State University of New York

Printed in the Unites States of America

For information, address State University of New York Press,
State University Plaza, Albany, N.Y. 12246

Production by M. R. Mulholland
Marketing by Nancy Farrell

Library of Congress Cataloging-in-Publication Data

Pusey, James Reeve, 1940–
    Lu Xun and evolution / James Reeve Pusey.
        p.   cm. — (SUNY series in philosophy and biology)
    Includes bibliographical references and index.
    ISBN 0-7914-3647-0 (hc : alk. paper). — ISBN 0-7914-3648-9 (pbk. : alk.
paper)
    1. Lu, Hsün, 1881–1936—Criticism and interpretation. 2. Lu, Hsün, 1881–
1936—Views on evolution. 3. Social Darwinism—China. I. Title. II. Series.
PL2754.S5Z8213 1998
895.1'85109—dc21   97-130 18
                                                                            CIP

10 9 8 7 6 5 4 3 2 1

To Qifang

∞

Contents

A Necessary Preface

"Lu Xun and Evolution" is a topic, as Wu Han, one of Lu Xun's ill-fated followers, said so often of other things, "worth our study today," a topic of "practical significance,"[1]—for those who care about China, and even for those, God forfend, who do not.

"Lu Xun and Evolution" is "worth our study today" for two reasons, one Chinese and political, one Darwinian and philosophical. For Chinese, "the two join and become one."[2]

One hundred and thirty-seven years after Charles Darwin wrote *On the Origin of Species*, we still do not know what to make of evolution. We do not know what to make of the *fact* of evolution, of that simple but amazing fact of life that present species have evolved out of former species, that present, ephemeral life forms have evolved out of former, ephemeral life forms, a phylogenetic fact of life, amazing but hardly more amazing than the ontogenetic fact of life (which we have recognized much longer) that we have individually evolved (though not in the same way) out of the meeting of an egg and a sperm, in a womb, on a rock, hurtling around a fire ball, in the middle of nowhere.

Darwin's predictions have come true. There has been "a considerable revolution in natural history,"[3] thanks to *The Origin of Species*. Whole new sciences have evolved. Darwin's book has led to libraries full of new knowledge about life.

And, of course, his theory has thrown "light . . . on the origin of man and his history"[4]—although we still know nothing of our ultimate origin.

That light has often been blinding.

It has not, at least, been wildly illuminating in the face of "the question of questions for mankind," not the question of "man's place in nature,"[5] but the question we must ask in the face of any natural fact: "So what?," the question Zhuang Zi's River Lord asked of Ruo, the North Sea God: "If that is so, then what should I do? What should I not do?"[6]

Darwin did not tell us. For a hundred and thirty-seven years people have sought in Darwin's theory of evolution "lessons for living," perhaps a great mistake. Lu Xun, with the rest, perhaps sought lessons that were not there.

Why, then, study "Lu Xun and Evolution?" Lu Xun (Zhou Shuren, 1881–1936), China's greatest twentieth century writer, was not a great scientist. He was not a great philosopher. He was a great critic of his people. He wrote to reform his people, or to get them to reform themselves. That was his cause, his only cause, and he wrote of evolution only for that cause. And yet his struggle to make something of evolution can help us in our struggle to make something of evolution.

Lu Xun's struggle to make something of evolution can also help us recognize a tragic source of modern Chinese suffering.

In the real world of Chinese history, and in the present world of Chinese politics, "Lu Xun and Evolution" is "worth our study today," because Darwin's theory of evolution deeply influenced modern Chinese thought—and still does; because Darwin's theory of evolution deeply influenced Lu Xun's thought; because Lu Xun is still the greatest "culture-hero" of the People's Republic; because Lu Xun influenced, and may still influence, the thought of millions of people in the People's Republic; because scholars in the People's Republic have written volumes of essays on "Lu Xun and Evolution"; because it matters how scholars and people in the People's Republic (and in China after the People's Republic) read, or will read, understand, and evaluate Lu Xun; because China's future still depends on Chinese answers to questions raised by Lu Xun about China's "evolution."

The theory of evolution deeply influenced modern Chinese thought because it was the first Western theory that most Chinese intellectuals thought true. Just exactly what was true about it was no clearer in China than anywhere else, but nonetheless, after 1895, after China had been humiliated in the Yellow Sea by the "solid ships and effective guns" of "Puny Japan," patriotic intellectuals came to believe that evolution was a fact of life they would have to take into account, a fact of life that would force itself on them whether they liked it or not, a fact of life they would have to live with or else.[7] This was clearly not because the theory of evolution was an interesting biological theory to be discussed over tea. It was not even because evolution offered startling new notions about the origin of species. Chinese paid attention to the theory of evolution because it seemed to make frightening sense out of China's predicament. It seemed to give dread warning that imperialism, the White Peril, was perfectly natural, a force against which it did no good to rail. History was natural history. The world was naturally a world of warring states and warring races. The Law of the Jungle was the only law. Only the fit would survive, and change was the secret of survival. So all started talking about change, and all who wanted change waved Darwin's banner: reformers and revolutionaries, constitutional Monarchists, Republicans, Anarchists, Nationalists, "May

Fourthists," Marxists. All waved Darwin's banner—and fought each other in their struggle for China's survival.[8]

Chinese "Social Darwinism" was strange Darwinism,[9] but it was clearly influential, and it influenced Lu Xun. He was just old enough at the end of the century to devour the translations and essays of China's first Social Darwinists: Yan Fu and Liang Qichao. He read *Tianyan Lun*, Yan Fu's paraphrastic translation of T. H. Huxley's *Evolution and Ethics*, reread it, and almost memorized it. And it stayed with him.[10] He went to Japan in 1902 to study mining, gave that up to study medicine, surely learned more biology than most of his contemporaries, but gave up medicine, and started writing.

He wrote, let us be clear, *not* to spread the gospel of evolution, but to save his people, to wake them up, to get them to change their ways, in thought and word and deed, to save themselves—from themselves. For Lu Xun, more than any other patriotic Chinese writer, was convinced that the Chinese were their own worst enemies.

But he wrote professing faith in evolution, as in nothing else, using Darwinian idiom in essay after essay, early and late—as, again let us be clear, did almost every other patriotic Chinese writer of the day.

But Lu Xun was not like any other writer of his day. He was the best Chinese writer of his day, and he remains the best twentieth century Chinese writer to this day. That is a fact, I think, beyond cavil. One does not have to like every word that Lu Xun wrote. One does not have to accept Chairman Mao's pronouncement that Lu Xun was "a great thinker and a great revolutionary" or that he was "New China's sage." But who can seriously contend with the Chairman's contention that Lu Xun was "a great writer?"[11] And who since, in China, has been greater?

Granted, Lu Xun was not a great writer when he wrote his first essays in *wenyan*, the literary language, from 1903 to 1908. His exaggerated, archaic, Zhang Binglin-like style marked a great leap backward from the liberated literary language of Liang Qichao, and a 180 degree bolt in the opposite direction from the language of the literary revolution that Lu Xun himself would finally champion. It was not until 1919, when he was pushed into the vernacular by editor friends of *Xin Qingnian* (the New Youth), that he "found himself" in the creation of a new language, a highly literary, iconoclastically erudite, powerfully subtle vernacular that no one since has used with such mastery.[12]

In this language, he created a startling new kind of short story, and in it he created the *zawen*, the "random essay" that was not a random essay, but a calculated, purposeful, focused—even when seemingly unfocused—terse tour de force of controlled, restrained, refined, and devastating sarcasm, hurled against, or lobbed at (over the heads of his censors) the maiming, murderous, or suicidal follies, as he saw them, of his people. His stories and essays antagonized many and inspired many. But in either case they made him, by his death in 1936, China's most famous writer.

And he still is, now more so than ever, thanks to the power of his work, to forces or accidents of history, and to the Propaganda Department of the People's Republic of China. Dying when he did, Lu Xun lent himself to the cause of the Chinese Communists, on whose side he had, indeed, at the end, taken his stand. Dead, of course, he could not leave it, and he has been held to it ever since, and not just by force. In the thirties his enemies, for the most part, were the Communists' enemies (though not exactly even then). He was at very least half converted to Marxism. Though never a Party member, he had Communist friends. And he placed what hope he could muster, although he could never muster much, on a Communist victory. So it was not strange that Mao Zedong should admire Lu Xun and claim him as a comrade, not strange that he should personally commemorate Lu Xun on the first anniversary of Lu Xun's death, in 1937, in an oration in which he called Lu Xun "the sage of New China." Soon enough, Mao Zedong would leave that oration out of his *Selected Works*, clearly convinced that he himself was "the Sage of New China," and beyond, but he demoted Lu Xun only slightly. He continued to extol him in vibrant strings of superlatives—that did win their way into the cannon: "On the cultural battle front, representing the great majority of all our people, charging the enemy and scattering their lines, Lu Xun is the most correct, most brave, most resolute, most loyal, most fervent, unprecedented national hero."[13]

So the Chairman himself made Lu Xun China's foremost culture hero. The Chairman himself set Lu Xun on the pedestal where the Propaganda Department has kept him—through all the twists and turns of the Party line and the Chairman's dialectic. In the People's Republic, Lu Xun has always been, officially, on the side of power (although not only there), and his books, therefore, have almost always been in the bookstores, where they have had a staying power matched only by that of the works of the Chairman himself, and the works of Marx, Engels, Lenin, and Stalin. Late in the Cultural Revolution, there even appeared in the bookstores a small, red book of quotations from Lu Xun, untitled, but the very same size as the other one.[14]

So Lu Xun has been in a unique position to influence the thinking of millions of people—and he has, albeit in diverse ways. In his lifetime he helped inspire many to rebel against "the great tradition,"[15] and against China's governments—warlord and Nationalist. In death he inspired some to look to the Communists. He helped inspire some, though not all, of his closest disciples to go to Yan'an, the Communists' stronghold in the barren mountains of Shaanxi. Patron saint of dissent, however, he also inspired some of his closest disciples, in Yan'an, as early as 1942, to criticize the Communists, loyally, they thought—and they paid for it, in lives of grief, lives that differed only in degree of grief.

Later, after the founding of the People's Republic, other followers also came to grief, like Wu Han, rash followers indeed. Yet, those who brought them to grief claimed themselves to be followers of Lu Xun—and their claims were not necessarily cynical. The Chairman, his wife, the rest of the Gang of Four, Lin Biao, Kang Sheng, Chen Boda, millions of "Red Guards," and millions of rank and file cadres, even "pragmatists" like Deng Xiaoping, all found occasion to quote Lu Xun. And all too many, people in power, or in pursuit of it, most horrendously in the Cultural Revolution, but also before and after it, all too many "used Lu Xun to beat people."[16]

That has had an ill effect on Lu Xun's influence. Some beaten by Lu Xun found comfort in Lu Xun.[17] But others, sick of the Gang of Four or Five's use of Lu Xun, turned against Lu Xun, or lost interest in him. In the early optimsim of the "China spring" that followed the dreadful winter of the Cultural Revolution, in the optimism of the decade of reform, before June 4, 1989, many simply found Lu Xun "too negative," " too sarcastic," "too destructive," "too unconstructive."[18]

Yet in that same "spring," officially supported scholars, not all supporters of officialdom, studied Lu Xun as never before. All over China research teams were organized, masses of research materials, research aids and research journals were published, as were scores of books, and thousands of articles—including droves on the topic of topics, "Lu Xun and Evolution."

Most Chinese scholars who have written on "Lu Xun and Evolution," however, have really written on "the Evolution of Lu Xun," taking as their nominal inspiration Qu Qiubai's 1933 thesis that Lu Xun evolved from an "evolutionist," to a Marxist, or "from the theory of evolution—to the theory of class."[19] Testing, and relishing, the new freedoms of Deng Xiaoping's "liberal" decade, however, from 1979 to 1989, some Chinese scholars at last dared find that formula too pat, though all feigned, at least, allegiance to its point: "Lu Xun evolved into one of us."

Echoing that refrain, all offered nominal support to the Communist Party. But at the same time some began, once again, with cautious courage, to praise Lu Xun as the patron saint of dissent.

Lu Xun would have rejoiced, impatiently, in his gradual liberation throughout that decade—until June 4, 1989, the day that confirmed his most pessimistic fears.

June 4 piled irony on irony. In the "spring" of 1981, one hundred years after Lu Xun's birth, a group had dared print on the program of a new stage production of *Ah Q zheng zhuan* (The true biography of Ah Q), under a glowering woodcut of Lu Xun, carved by one Wang Weixin ("Reformer Wang"), lines that Lu Xun had meant to be ironic in 1926:

> I would be very happy if it were true, as people say, that I have only written about former times or a specific period. But I am still afraid that what I have seen may not be what preceded the present but perhaps what will come hereafter, maybe even twenty, or thirty years hereafter.[20]

Or sixty. The simple irony is that Lu Xun still speaks specifically to China's present in a "future" twice as old as the future he feared. (On another level, of course, he speaks, and will continue to speak, to the present of us all).

But there is a sadder irony. As the young Lu Xun scholar, Wang Hui, bluntly stated at last, in 1988, Chinese Lu Xun scholars have helped hinder Lu Xun from speaking to the present, even when arguing that he should.

Nowhere, alas, has this been clearer than in what Chinese scholars have written about "Lu Xun and Evolution," for they have unwittingly propogated and perpetuated a myth about evolution—even more than a myth about Lu Xun—that has ironically perpetuated the seemingly perpetual present that Lu Xun so fervently longed to destroy.

But Lu Xun himself helped propogate and perpetuate that myth—saddest irony of them all. That is why he was on both sides of Tiananmen Square, on June 4, 1989.

There is a myth about evolution, a superstition, starkly visible in the tragedy of June 4, 1989, that Lu Xun himself did not see through, that the world must see through, before "the banquet of human flesh" that Lu Xun so well decried will ever end.[21]

Caveat lector. This book is not an ordinary monograph in Chinese intellectual history. It is not just about China. It is not just about Lu Xun. It is certainly not an introduction to Lu Xun, or to his works. It is not an intellectual biography. It is not "an appreciation." It is not a study of Lu Xun's genius or his art (although both will shine through). It is a philosophical critique of Lu Xun's thought and a philosophical and political critique of what Chinese in the People's Republic have done, and may yet do, with Lu Xun's thought, *and* it is a reflection on philosophy and biology.

This is a book about Lu Xun *and* evolution, because I am convinced of three things, that one cannot understand Lu Xun's thought without thinking hard about Lu Xun and evolution, that one cannot do that without thinking hard about *evolution* (One of the great weaknesses, it seems to me, of most, not all, of the masses of Chinese studies of Lu Xun and evolution lies in the assumption, *not* just Chinese, that evolution is an open book—an assumption with very real, and very sad, consequences), and that one cannot pursue *veritas* and the Way without taking evolution into account.

I am not at all tempted to shout "Amen," when Edward O. Wilson proclaims that "scientists and humanists should consider together the possibility that the time has come for ethics to be removed temporarily from the hands of the philosophers and biologicized."[22] But I do applaud Huxley, who said in 1892: "It may well be, that it is only my long occupation with biological matters that leads me to feel safer among them than anywhere else. Be that as it may, I take my stand on the facts of embryology and of palaeontology; and I hold that our present knowledge of these facts is sufficiently thorough and extensive to justify the assertion that all future philosophical and [an odd addendum for Huxley] theological speculations will have to accommodate themselves to some such common body of extablished truths as the following"—twelve truths about evolution that Huxley held to be self evident.[23]

Huxley tried to accomodate his thought to evolution. Lu Xun tried. We must try.

But Huxley and Lu Xun failed.

And in all honesty, if to accommodate our thought to evolution we must make sense of it all, we will fail too.

And yet—light will be thrown on the nature of our existence in our very failure.

That is why Lu Xun and evolution is a topic worth our study today.

Acknowledgments

At long last I can acknowledge in print the help of six institutions that have probably long since written off their help to me as unproductive:

The Committee on Scholarly Communication with the People's Republic of China gave me a grant to do research in China during 1981–1982, and most graciously renewed it for an additional half year.

The Literature Institute of the Chinese Academy of Social Sciences, my host unit for the first year, took excellent care of my family and me, introduced me to a splendid host of Lu Xun scholars, and gave me, or helped me find, mountains of research materials.

The Philosophy Department of Beijing University took over as my host unit for the next half year, giving me an unforgettable term on that most exciting of Chinese campuses.

The Wang Institute of Graduate Studies gave me a Fellowship in Chinese Studies four years later, and, through it, a whole year free for writing, a boon indeed, that cut years off the time it would otherwise have taken me to finish this book.

The American Council of Learned Societies gave me a travel grant in October 1986 that let me accept an invitation from my pervious benefactors at the Literature Institute of the Chinese Academy of Social Sciences to attend an intensely exciting conference in Beijing, marking the 50th anniversary of Lu Xun's death, where I presented a first outline of this book's thesis.

Bucknell University not only granted me leaves, sabbatical and otherwise, that let me accept the above assistance, but also gave me a grant for my index.

I thank all those institutions. But institutions do not really exist. I really thank people, known (too many to name—but not forgotten) and unknown.

Some of those people, however, in China, will feel, I fear, that in the pages that follow I have bitten hands that fed me (both with intellectual fare and with Chinese feasts)—and I have. But *que faire?* I set out to see what I could see, and this is what I saw. If some now wish they had not helped me, I can only offer them thanks and apologies.

I must apologetically thank, also, Chinese friends whom I dare not name, friends who helped me greatly in my research but who might still fear, I fear, the dangers of guilt through association with a foreigner whose conclusions will be anathema to the Party faithful (an endangered species). I hope my caution is silly, but I cannot yet quite be sure.

Those friends are in no way whatsoever, of course, to blame for my conclusions. Nor are those who most generously wrote letters on my behalf, urging institutions to gamble on my project, Chou Chih-p'ing, Nicholas Clifford, Richard Ellis, Jerome Grieder, Perry Link, Benjamin Schwartz, and Richard W. Wilson. Nor are those who with sharp intelligence and eyesight read and commented on my manuscript, Richard Ellis, Kenneth J. Hsu, Jan Pearson, Hoyt Tillman, Susan Warner-Mills, and two anonymous readers, all of whom gave me most perspicacious advice—most of which I took.

I must acknowledge, however, my failure to take two major suggestions: I have failed to seek help from Japanese scholars—because my Japanese is vestigial, and I have failed to analyze the differences between my view of Lu Xun and the views of prominent American scholars—because I could find no nondisruptive way to fit such an analysis in, either in the middle of my argument, or in the preface (before I had made my argument), or at the end, and because I am not yet sure myself how deep our differences are. In this book I have concentrated on Chinese scholarship, not Amnerican or Japanese, because Chinese scholarship is part of "the story," the ongoing Chinese story.

For turning my study of that story physically into a book, I am deeply grateful first and foremost to Joyce Miller, who with great skill and patience "processed" my entire pencil-scribbled manuscript. I am grateful to troops from Bucknell's Computer Center, most especially to three students, Dustin Torsello, and two others who's names I cannot recover, who saved at the last my edited disks from the best destructive efforts of sinister viruses—and me.

Again I owe a tremendous debt to Florence Trefethen, editor of my last book, saving index compiler, sans pareil, of this one.

I am also, of course, supremely grateful to all who have worked on this project at the SUNY Press, first to Professor David Edward Shaner, Editor of SUNY's Philosophy and Biology Series, for accepting it in his series, then to William D. Eastman and Nancy Ellegate, to my most helpful Production Editor, Megeen R. Mulholland, and to my able Copy Editor, Laura Glenn, as well as the typesetter and printer.

For most gracious permission to quote from wonderful songs, I am indebted to Folk Legacy Records, in Sharon Connecticut, for a few choice words from "Freedom on the Wallaby," words by Henry Lawson, sung by the inimitable Gordon Bok, and to Mrs. Claudia Flanders, for lines from "The Reluctant Cannibal" and "Dead Ducks," written by her equally inimitable late husband, Michael Flanders, and sung to the music of his just as inimitable colleague, the late Donald Swann.

Finally I acknowledge the wonderful support of kith and kin, who have brought me great cheer throughout the whole slow course of my study of Lu Xun and evolution. I give heartfelt thanks to them all, most especially to my parents and my mother-in-law, who, once again, have waited so patiently for me for so many years, to my children Drake and Jennifer, who have grown up with this book, and put up with the uncountable hours I lost myself at my desk, and to my noble wife, Qifang, who has put up with most of all, and given most of all, and to whom, in loving thanks for all, this book is dedicated.

1

A Mentor Once Removed

In the beginning Lu Xun read *Tianyan lun. The Origin of Species* was not the origin of his Darwinism. He discovered Darwin in Yan Fu's translation of T. H. Huxley's *Evolution and Ethics*, on a long lost Sunday afternoon sometime in 1901, when he was a twenty-year-old student in a late-in-the-Dynasty's-day "self-strengthening" School of Mining and Railways in Nanjing.[1]

1901 was a good year for beginnings, new beginnings, or at least new dreams, for the year before had been as disheartening a year as China had seen in three hundred. Of course, in ordinary human terms—and what other are there?—the tragedy of the midnineteenth century Taiping Rebellion was a thousand times worse than that of the Boxer Rebellion of 1900, but any patriot who had watched China's defeat in the Sino-Japanese War of 1895, the imperialists' scramble for concessions after it, and the failure of the Reform Movement of 1898 after that, must have thought that the miserable unfitness of the Boxers and the murderous arrogance of the avenging armies of seven Western imperialist nations, including, alas, the United States, and of one Eastern apprentice-imperialist nation, Japan (a quick learner), meant the end.[2]

But 1900 was not the end. The barbarian troops gave Beijing back, although they did not all leave. The "Powers" did not "carve up the Chinese melon," although they marked out spheres of influence. They let the Empress Dowager come back, with her unfortunate nephew, the Emperor, in tow, and let her put him back in his luxurious palace island house arrest. The Manchu Dynasty survived. But in 1901, when Lu Xun discovered Darwin, the Dynasty did not look fit.

And yet on that long lost Sunday, Lu Xun seems momentarily to have forgotten the Dynasty and the Empress Dowager and the Barbarians and the Boxers. In his description of that day, years later, we can still sense a transcendent excitement:

So it became popular to read new books, and I learned that China had a book called *Tianyan lun*. On Sunday I ran down to the southern section of the city and bought a copy, a thick lithographed copy on white paper, for exactly five hundred wen. I opened it and took a look. It was written in excellent characters, and the first lines read: "Huxley sat alone in his house in southern England, with mountains behind him and fields before. The scenery outside his windows was as clear as if at his fingertips, and he wondered what had been there two thousand years ago, before even Rome's great general, Caesar, had arrived. And he guessed that there had been only wilderness, created by nature." —Oh, so the world has a Huxley, thinking like that in his study, and thinking so freshly. I read on without stopping, and I came upon "the struggle for existence" and "natural selection," and I came upon Socrates, and Plato, and the stoics.[3]

He discovered Darwin, and Spencer, and Huxley himself all at once, and he discovered much more: Homer, Hamlet, Kant, and Hume, Shakespeare, Thales, Haeckel, Job, Alexander the Great, Alexander Pope, cynics, tudors, chimpanzees, biology, logic, nebulae, nerves. He discovered whole new worlds.

As a boy Lu Xun had by nature been fascinated by nature. In his mining school he had encountered with excitement the rudiments of physics, chemistry, and geology. But it was in *Tianyan lun* that he discovered the interrelated worlds of Western science, Western literature, and Western thought. He never closed his mind to these worlds, though he may have closed his mind to some things in them. He never forgot the book. Years later he could still recite whole passages from it by heart.[4]

But what was it in *Tianyan lun* that so excited him? In *Tianyan lun* there were many voices, Darwin's and Huxley's and voices Huxley sought to refute. In Yan Fu's commentary one could hear Herbert Spencer *über alles*, but also Yan Fu himself. Whom did Lu Xun hear, and did he hear correctly? That is what we must ask of Lu Xun's first essays, written in Japan in 1903, and of his second batch of essays, written still in Japan in 1907 and 1908. But first we must say *something* of the book behind *Tianyan lun*, of *Evolution and Ethics*, which Lu Xun did not read.

Evolution and Ethics, T. H. Huxley's Romanes lecture of 1893, and its "Prolegomena" of slightly greater length, which he wrote afterward to try to say another way what he had tried to say the first time, is too good a book to be paraphrased in a paragraph, or two, or three. It should be read. It is, and will always be, worth reading. But we must say something about it before we begin, because it is a book we must bear in mind throughout this book. Its title, which Yan Fu, alas, chose not to translate, holds the hidden question behind our title, and gives us our true topic, a topic of academic interest to outsiders, of much deeper import to Chinese: Lu Xun, evolution, and ethics.[5]

Huxley wanted evolution *and* ethics. He wanted his evolution and his ethics too. He believed in evolution absolutely. He was "Darwin's bulldog" to the end—ever, if necessary, "episcopophagous."[6] But when he went to Oxford in 1893, to deliver what would turn out to be virtually his last will and testament, back where he had made a monkey of Bishop Wilberforce thirty-three years before, he went not to attack any more benighted bishops (a subspecies, c.f. Gongsun Long: "A white horse is not a horse."[7]) opposed to evolution, but benighted evolutionists opposed to (Judeo-Christian) ethics, simpleminded might-is-rightists ready and all too willing to do their worst, and ours, in Darwin's name.

Huxley was disgusted by Social Darwinists who preached "fanatical individualism," practiced "reasoned savagery," given the chance, and glorified, as long as they were winning, "the gladiatorial theory of existence."[8]

He was disgusted by Social Darwinian "pigeon fanciers" who spoke of bettering our breed by weeding out the unfit—or at least by letting them die ("If they are sufficiently complete to live, they do live, and it is well they should live. If they are not sufficiently complete to live, they die, and it is best they should die,"—for "under the natural order of things society is constantly excreting its unhealthy, imbecile, slow, vacillating, faithless members"[9]).

And he was especially disgusted by Social Darwinists who hid such ruthless indifference behind paeans of praise for evolution inexorably working its way, in Darwin's own phrase, alas, "towards perfection."[10] Huxley believed in no such perfection, certainly none won over the dead bodies of the unfit (though who could deny that that is how our species won its existence?). "The theory of evolution encourages no millennial anticipations," he said:[11] "The prospect of attaining untroubled happiness, or of a state which can, even remotely, deserve the title of perfection, appears to me to be as misleading an illusion as ever was dangled before the eyes of poor humanity. And there have been many of them."[12] Huxley was disgusted by those whose sugar-plum visions of perfection blinded them to the price of evolution, to its pain.[13] He was disgusted by those who blithely said that "all is well since all grows better."[14] He did not believe that all grows better, and he was sure that all was not well:

> I know no study which is so unutterably saddening as that of the evolution of humanity, as it is set forth in the annals of history. Out of the darkness of prehistoric ages man emerges with the marks of his lowly origin strong upon him. He is a brute, only more intelligent than the other brutes, a blind prey to impulses, which as often as not lead him to destruction; a victim to endless illusions, which

make his mental existence a terror and a burden, and fill his physical life with barren toil and battle.[15]

"If anything is real," said Huxley, "pain, and sorrow, and wrong are realities." Evil is real. And "evil stares us in the face on all sides." It does, it always has, and it ever shall.[16]

But what then should we do? Leap into "the great work of helping one another":

> I think I do not err in assuming that however diverse their view on philosophical and religious matters, most men are agreed that the proportion of good and evil in life may be very sensibly affected by human action. I never heard anybody doubt that the evil may be thus increased, or diminished; and it would seem to follow that good must be similarly susceptible of addition or subtraction. Finally, to my knowledge, nobody professes to doubt that so far forth as we possess a power of bettering things, it is our paramount duty to use it and to train all our intellect and energy to this supreme service of our kind.[17]

Huxley was for human action. That is why the Chinese liked him. He was disgusted by those who said evolution tells us we can do nothing ("You and I can do nothing at all. It's all a matter of evolution. We can only wait for evolution. Perhaps in four or five thousand years evolution may have carried men beyond this state of things").[18] But he was equally disgusted by those who said it tells us we can do anything ("All's fair in love and war" and in "the interneccine struggle for existence").[19] That was, he admitted, what evolution tells us. It tells us we can do anything at all that will fit us to survive, and to be fruitful and multiply. But we should not listen to it. Its message is not for us. At least it is no longer for us. It is a great mistake "to apply the analogy of cosmic nature to society." Rabid social Darwinian lawyers from the Law School of the Jungle are guilty of yet another "misapplication of the stoical injunction to follow nature."[20] They fail to see "that cosmic nature is no school of virtue, but the headquarters of the enemy of ethical nature."[21] They fail to see that "social progress means a checking of the cosmic process at every step and the substitution for it of another, which may be called the ethical process."[22] Granted, "for his successful progress, throughout the savage state, man has been largely indebted to those qualities which he shares with the ape and tiger," but now, "after the manner of successful persons, civilized man would gladly kick down the ladder by which he has climbed. He would be only too pleased to see 'the ape and tiger die,'" for "there is a general consensus that the ape and tiger methods of the struggle for existence are not reconcilable with sound ethical principles."[23]

Man's way was the ethical way. Huxley was a good Confucian. But there was something wrong, if not with his rhyme, with his reason. At least there was something very strange.

There was something stranger than ever about "man's place in nature." We had been in an odd enough position at the end of Huxley's first book, *Man's Place in Nature*, which he had written in 1860 to defend Darwin and to fulfill the promise of the one sentence Darwin had written in *The Origin* about our origin. "Light will be thrown on the origin of man and his history."[24] Huxley *did* throw light on what he called "the question of questions for mankind—the ascertainment of the place which Man occupies in nature and of his relations to the universe of things. Whence our race has come; what are the limits of our power over nature, and of nature's power over us; to what goal are we tending. . . ."[25] At least he threw light on the question of whence our race has come. He produced respectable scientific evidence to support the thesis that everyone knew Darwin had implied: that we are flesh and blood relatives of "the brutes."

But even as Huxley the scientist argued that point, he argued in a most unscientific manner about "the grandeur of the place Man occupies" in nature.[26] "No one," he said, "is more strongly convinced than I am of the vastness of the gulf between civilized man and the brutes; or is more certain that whether *from* them or not, he is assuredly not *of* them."[27] Thanks to man's "marvelous endowment of intelligible and rational speech," which has allowed him alone to accumulate and organize his experience, man now stands "as on a mountain top, far above the level of his humble fellows, and transfigured from his grosser nature by reflecting, here and there, a ray from the infinite source of truth."[28] With our feet in the clay and our heads in the clouds, what *was* our place in nature? Huxley, in a clairvoyant, or lax, moment had waxed as mystic as Darwin had in his famous lax or clairvoyant moment in the primeval forests of Brazil. But in his autobiography, Darwin, declaring himself an agnostic, recanted his Brazilian faith that "no one can stand in these solitudes unmoved, and not feel that there is more in man than the mere breath of his body."[29] What is odd is that Huxley, the father of agnosticism, never saw any need to recant his metaphysical pronouncements about rays reflected "from the infinite source of truth."

Instead, in *Evolution and Ethics*, he pushed us up to an even higher peak. He had us not only "from but not of" the brutes but "from but not of" evolution. He raised us high enough above evolution to condemn it. "Brought before the tribunal of ethics," he said, "the cosmos might well seem to stand condemned. The conscience of man revolted against the moral indifference of nature, and the microcosmic atom should have found the illimitable macrocosm guilty."[30] Huxley was revolted by the "moral indifference of nature," and he called on us to revolt against it, to combat it "at every step." But how could we? Who were we to judge the universe? What were we that we could rebel against Mother Nature that begat us? This was filial impiety of truly cosmic proportions. It also seemed logically impossible. For Huxley himself had done his best to prove us perfectly natural. He had argued that "man, physical, intellectual, and moral, is as much a part of nature as purely a product of the cosmic process as the humblest weed."[31] He had said that those who argued "in favor of the origin of moral sentiments in the same way as other natural phenomena, by a process of evolution" were "on the right track."[32] He seemed to believe in that view that so pained Darwin's wife, Darwin's own opinion "that *all* morality has grown up by evolution [italics Mrs. Darwin's]."[33] But still he pitted the ethical process *against* the cosmic process, protesting "that if the conclusion that the two are antagonistic is logically absurd, I am sorry for logic, because, as we have seen, the fact is so."[34] But how could it be so? In sicking us on evolution was he not setting evolution at its own throat?

Of all the creatures of evolution, man was odd man out. And Huxley said he should be. He said we should refuse to go along. That in itself meant that our place in nature was a lonely one, exalted or no. But it was even lonelier if one accepted Huxley's protestation that our cosmic resistance was doomed to defeat. Remember, Huxley had no "millennial anticipations."[35] He had no faith that man would live happily or at all ever after, in this world or any other. At best all we could do was cultivate our garden until in "the procession of the great year," "the evolution of our globe shall have entered so far upon its downward course that the cosmic process resumes its sway; and, once more, the State of Nature prevails over the surface of our planet."[36]

Now granted, Pogonian fears about the sun expanding in ten million years killing all life ("and me so young")[37] may seem silly. Less silly is the certainty Huxley was blessed not to know, that we could blow ourselves off our planet in the twentieth century. But philosophically the fact remains: Huxley granted man no immortality even as a race. He placed man, all mankind, as a *species*, in a natural existential predicament. "Out of the darkness," he said, we come.[38] And unto it we shall return, unto the darkness of biological extinction.

Staring evolution in the face, Huxley came to the existential question: "What do we do in the meantime?" And the answer for Huxley was perfectly simple: "Choose to act ethically." Why? Because (*pace, feminae*):

We are grown men, and must play the man

strong in will
to strive, to seek, to find, and not to yield,

cherishing the good that falls in our way, and bearing the evil in and around us, with stout hearts set on diminishing it. So far, we all may strive in one faith towards one hope:

> It may be that the gulfs will wash us down,
> It may be we shall touch the Happy Isles,
> . . . .but something ere the end,
> Some work of noble note may yet be done.[39]

Again, noble rhyme, half of it Tennyson's, but odd reason—for an agnostic, proto-existentialist. "*Naturam expellas furca tamen usque recurret.*"[40] Nature would win, but we must strive against it, and against evil. And somehow, as Camus would later say, "The struggle itself toward the heights is enough to fill a man's heart. One must imagine Sisyphus happy."[41] That true existentialist "creed" was strange enough, but Huxley's was stranger: "The practice of self-restraint and renunciation," necessary for our struggle, "is not happiness," he said, "though it may be something much better."[42] But what in Heaven's name was that? Why, even if we did possess "a power of bettering things," was it "our paramount duty to use it?"[43] Whence came that duty? Why should we do "work of noble note," and how were we to know what it was? How did Huxley know? That was the strangest thing of all. The agnostic knew. "That one should rejoice in the good man, forgive the bad man, and pity and help all men to the best of one's ability is surely," he said, "indisputable. It is the glory of Judaism and of Christianity to have proclaimed this truth, through all their aberations."[44] How did he know? Who knows? Not knowing how he knew, he said through "intuition."[45] But intuition was a strange power for "a product of the cosmic process" as natural "as the humblest weed," and it was illusion if there was nothing to intuit.

What does this all prove? It is not clear. But clearly Huxley, *episcopophagous*, who had no use for any god, who would not believe in one without "good grounds for belief" and who saw none, believed in good, a good above and beyond the good-for-me-and-my-genes good of evolution.[46] What his belief proves is indeed unclear, but we must bear it in mind. When we begin to hear modern scholars praise *Tianyan lun* for starting Lu Xun on his inexorable march toward materialism, we must remember how much Huxley, his mentor once removed, despite his own best efforts, was an idealist.

2

The Pen, Not the Scalpel or the Sword

When he was twenty, Lu Xun went to Japan. Landing in Yokohama on April 4, 1902, with a small band of fellow graduates of the Nanjing School of Mining and Railways, chosen, sent, and supported by the Manchu government, he was expected to learn Japanese, study more about mining, and then return to serve his country—and his dynasty. For two years he did what he was supposed to, more or less. At least he diligently studied Japanese, and German, and a little science, in a special Tokyo preparatory school for Chinese. But then, instead of trying to fight his way into Tokyo Imperial University to study mining in earnest, he took the advice of one of his Japanese teachers and went off alone to Sendai to study medicine.[1]

In those first two years, however, he set out, without realizing it, on his true career. He started to write. He wrote and published his first four bonafide articles. He published his first two translations. And he flirted with revolution. But he did not distinguish himself as a great writer, a great thinker, or a great revolutionary.[2] Indeed it is hard to find even hints of greatness in either his thought or action during those first two years, although some scholars have managed to do so more easily than others.[3] His writing was unrepentantly old fashioned, his thinking was unremarkably typical of that of other patriotic students in Japan, and his "revolutionary acts," though not without some risk to his government stipend, were tame.

And yet we can glimpse concerns in Lu Xun's early thought and action that would stay with him, and prove important, however unremarkable in their beginnings. Our first record of "the thought of Lu Xun," for example, happily for us Darwinian, is important precisely because it was so common: Lu Xun's first intellectual concern when he went to Japan was China's fitness—or rather unfitness. That is what he and his best friend Xu Shoushang talked about almost from the first day they met in September 1902. As Xu Shoushang would later recall in at least two different reminiscences:

One day we were talking about how terribly cheap Chinese lives were
in history, especially when Chinese were enslaved by foreign races,
and we could only face each other in sad silence. Thereafter we became even closer friends, and whenever we met, we talked about the
failings of the Chinese national character. Because we were in a foreign country there were many things to upset us. . . .We also often
talked about three related questions: 1) What was mankind's ideal
nature? 2) What did the Chinese race lack most? 3) Where lay the
source of its sickness?[4]

That was the pressing question. What was the source of China's
sickness, China's weakness, China's unfitness? Every patriotic student
knew of the insulting Englishman who had branded China "the invalid of
the Orient," and every patriotic student knew by now the frightening Darwinian context into which the would-be reformers Yan Fu and Liang
Qichao had put that insult: In the jungle world of warring states, sick nations, weak nations, unfit nations perished in the natural international
struggle for existence.[5] Whole peoples perished—where there was no vision. But vision could make a people fit. That was the Lamarckian
promise—although called Darwinian—with which Liang Qichao had
countered his own vision of Darwinian doom. Armed with a fortuitous
mistranslation and a misunderstanding clearly based on wishful thinking,
Liang Qichao had proclaimed that "Mr. Darwin has said every living
thing, no matter of what kind, must frequently change its form and make
it beneficial to itself, for only thus may it survive."[6] The Lamarckian promise was that one could change one's form. One could make oneself fit.

But what Darwin had actually said was that "any being, if it vary
however slightly in any manner profitable to itself, under the complex and
sometimes varying conditions of life, will have a better chance of surviv-
ing, and thus be *naturally selected*."[7] There was a world of difference
between a being or species that could "change its form and make it beneficial to itself" and one that might "vary" in a "manner profitable to
itself—and thus be *naturally selected*," but Liang Qichao, overlooking the
world of difference between transitive and intransitive verbs, leapt to the
conclusion that the Darwinian secret of survival for human beings was not
natural selection but selection by man. "Selection by man," he said,
"means carefully to seek out the unfit in oneself and change it, to make
oneself fit to survive."[8] And "selection," wrote the great would-be reformer, "is revolution."[9]

Liang Qichao leapt to Lamarckian conclusions. So did most of his
readers. So did Lu Xun. And they all leapt to potentially revolutionary
conclusions. There was revolutionary potential in Lu Xun's search for the

source of China's illness. But there was nothing remarkable about his search. Every patriot with any vision at all was searching for the source of China's seeming unfitness. Liang Qichao's "Darwinian" faith—which Lu Xun shared—that the source of that unfitness could be recognized and rooted out, did not make evolutionary sense. No biological "creature" had ever fit itself to survive. True, eugenic fiddling may now be possible, but eugenic fiddling had nothing at all to do with the evolution that has made eugenic fiddling possible. Evolution had not "progressed" through the efforts of organisms to change themselves.

And yet if Liang Qichao's novel Lamarckian metaphor did not make scientific sense, Lu Xun's common medical metaphor did make common sense. Who could doubt it? What could be more "natural" then to diagnose one's illness, cure it, and so strengthen oneself, the better to fight the good fight? That is what Chinese had been trying to do ever since 1860. What was new was that now even the reformer Liang Qichao recommended a "revolutionary" cure, although, eschewing violence, the "revolution" he had in mind was still reformist.[10] At any rate, small wonder that Xu Shoushang would later claim that he and Lu Xun concluded at the time that "the only cure [for China's ills] was revolution."[11]

And yet Lu Xun's first essays were not revolutionary. His first effort might have been. "Sibada zhi hun" (the soul of Sparta) was a weapon-waving piece that extolled in stilted, melodramatic classical Chinese the Greek version of that later classic Roman refrain, *"Dulce et decorum est pro patria mori."* Lu Xun retold the tale of the Battle of Thermopylae, building in extravagantly archaic prose a monument to the wife of the lone survivor, the unfortunate chap with eye trouble, who slipped away to fight again another day, rather than blindly die without striking a single well-aimed blow for his fatherland. He went home to profess once more his love for his wife, while he waited for a better time to give his life. But she, ashamed, cursed him for his "cowardice" and killed herself to spur him back to a death now doubly owed his state. He paid that debt at the Battle of Plataea, and his general granted that at the end he showed a Spartan soul. But because he had survived a defeat, and so had broken Spartan law, he was granted no burial. Instead Sparta raised a monument to his wife.[12] And Lu Xun did likewise.

But Lu Xun, faithful indeed to the soul of Sparta, praised militant self-sacrifice in the name of national defense not revolution. His cry was for Chinese to fight with Spartan spirit not against their monarch or the Manchus, but against the *Russians*. For like all Chinese students in Japan in 1903, Lu Xun was up in arms—figuratively—because the Russians were threatening to seize Manchuria. He was only one of a host of patriotic polemicists, seasoned and unseasoned, who turned to Sparta for inspiration.

Who first discovered Sparta it is hard to say—probably some Japanese. But it was not Lu Xun. Liang Qichao urged Chinese to be Spartan, militantly Spartan, half a year before Lu Xun did.[13] And a month before Lu Xun told or translated his tale of Thermopylae, a band of Chinese students in Japan used the example of that battle in their open letter to the Qing Court, demanding that they be organized into a "volunteer army to resist the Russians." (They were not.) "Unto this very day," they wrote, "the glorious fame of the Battle of Thermopylae shakes the nations of the world. In the West every three foot child [sic] knows the story. Just think. If even in that little Greek peninsula there were righteous warriors who would not disgrace their nation, in the millions of square miles of our empire can there be none?"[14] That was the question Lu Xun echoed, though his echo was a bit odd in a story whose hero was its heroine: "Alas are there no men today unwilling to be worse than women? There must be some who will throw down their pens and stand up?"[15]

Perhaps there were, but Lu Xun was not one of them. Nor would he ever be. He would never *"tou bi cong rong"* (throw away his pen and join the army). For all his brave talk (and some of it was brave) his pen would be his only sword—or spear, or dagger.

But if he would not hearken to his own Spartan call to arms, of what import was it? Not much, except that it revealed a hard, callous, unfeeling, nationalistic streak in him that we must keep in mind. The future crusader against customs, laws, and morals, in his own country, that "ate people," extolled in "the Soul of Sparta" as cannibalistic a national ethos as the world has ever known. The Spartan People devoured Spartan people. Spartan eugenics, Spartan martial law, Spartan "honor," all ate people. Lu Xun extolled a Spartan law that demanded senseless self-sacrifice. He extolled a Spartan sense of shame that led to senseless suicide. In 1919, his "madman" would see in the words "righteousness and morality," scrawled over every page of Chinese history, the words "eat people."[16] Why in 1903 could he not see the same words in the pages of Spartan history?

There is another thing we should keep in mind. In "The Soul of Sparta" Lu Xun applauded the total subjugation of romantic love to love of country. Applauding the notion that "a Spartan warrior speaks of no love above his country," Lu Xun showed no more sympathy for the warrior who returned to see his wife once more before he gave his life, because he loved her, than the warrior's Spartan wife did:[17] *Ai guo* (love of country) *über alles*. Was it complete coincidence that three years later Lu Xun should briefly return to China and seemingly sacrifice all hope of romantic love himself by accepting a loveless marriage arranged for him by his mother? Was it complete coincidence that a few days after his wedding he should return alone to his duty in Japan, sacrificing on the twin altars of *zhung* and *xiao* (loyalty and filial piety) a bride doomed to a life of cool support and cold neglect, a bride he would feed but also starve? Lu Xun's wedding feast was a perfect example of the "cannibalism" he would later decry.[18] When he finally did decry it, he would know of what he spoke.

At that time filial piety was the more conspicuous of the altars on which he sacrificed his bride's happiness—and his own.[19] But one senses something of "The Soul of Sparta" in his *mamu buren* (unfeeling and inhumane) behavior. Thirty-one years later he would write that on rereading that essay he could not prevent his ears from burning—with shame. But he seems to have been more embarrassed by his essay's style than by its content. "Although I am ashamed of the works of my youth," he wrote, "I do not regret them. Indeed I am still quite fond of them."[20] — And so two years before he died he reprinted them. We must remember that, at the end, when we ask whether his ears burned with reason.

At any rate, there was precious little that was revolutionary in "The Soul of Sparta." Nor was there much that was revolutionary in his next works, in his first translations and in his essays on radium and on China's mineral resources.

Lu Xun's first translation, of an odd little reminiscence about Victor Hugo, written by Victor Hugo's wife (although falsely attributed by most Chinese scholars—and by Lu Xun—to Victor Hugo himself) was a restrained protest against official injustice (being a true account of Victor Hugo rescuing a poor young woman from a six-year jail sentence summarily given her for pummeling a young dandy who put a snowball down her back), but there was no indication that Lu Xun looked to revolution rather than to reform to correct such injustice.[21]

The rest of his translations of 1903 were even less revolutionary, for they were translations, or a translation and a half, of two "science fiction" novels of Jules Verne, which Lu Xun translated to excite interest in real science, and in the study of it.[22] Real works of science, said Lu Xun, too easily put readers to sleep. But Jules Verne's *De La Terre à la Lune* would awaken Chinese to an all important fact of life: "Man is an animal with hope of progress"—the alleged promise of evolution.[23] And science fiction would awaken Chinese to the fact that science could save the nation. But that was the once and future cry of the gradualist reformers Yan Fu and Liang Qichao, and in time Hu Shi and even Deng Xiaoping. It was not a cry for revolution—however revolutionary its consequences.

Lu Xun's report, "Shuo ri" (on radium), was a perfect example of a story of scientific fact that could put people to sleep, even as it conveyed something of the excitement of step-by-step scientific progress. With archaic grammar and a bewildering host of neologisms, Lu Xun told the story of Mme. Curie's recent discovery of radium. From somewhere he managed to marshal an impressive array of facts about the mysteries of radioactivity, although he did not display any deep understanding of those facts. Years later he would confess that his scientific level had at the time been low, and that he had "stolen" his information, but at least he had honestly expressed wonder at the wondrous ways scientists unwittingly worked together to make progress. "From the study of X rays," he wrote, "came the discovery of radium emanations; from the study of radium emanations came the theory of electrons and from that, in a flash, came a great change in our concept of matter." And so, he said, "although we recognize Mme. Curie's great contribution, we should really take our hats off to Mr. Röntgen, the late nineteenth century discoverer of X rays."[24] In the world of science, revolutionary discoveries were evolutionary.

What was the point? It is hard to say. Wang Shiqing in a recent commentary has said that even today "this paper, in dispelling superstition, liberating thought, propagating materialism, and opposing metaphysics, offers us great inspiration."[25] But inspiration lies in the eye of the beholder. It is not easy to find inspirational preaching in the original. One can see a Westernized, pragmatic, down-to-earth dedication to the good old neo-Confucian investigation of things and the extension of knowledge, but if that was made novel with its neologies, it was still not inspirationally revolutionary—because Lu Xun came to no revolutionary conclusions.

Only in his essay "Zhongguo dizhi luelun" (a brief discussion of the geology of China) were there hints of revolutionary ire, hints that would somewhat compromise his one and only effort to repay his government's six-year investment in his study of mining—before he gave that up to study medicine. Three years later he would publish, with a classmate, one echo of his essay—but thereafter he would never mention mining again.[26]

With or without hints of revolutionary ire, however, his all too brief discussion of Chinese geology could hardly have seemed sufficient repayment for six year's support, even though his basic thesis was one the Qing government applauded. For, after all, all he argued was that China should seize control of its resources before foreigners did.

But, alas, to prove his point he had to rely, rather ominously, on those very foreigners' own reports on the extent of China's resources— most especially on the extent of China's coal reserves. He tried to cheer his readers up by citing the contention of a German geologist, F. von Richthofen, that China was "the number one coal country in the world," but he also purposely frightened his readers by citing the worries of many that coal would do China in, by inviting imperialist aggression.[27] One way or another he said, "Coal is intimately related to a nation's economic growth or decline. Indeed it is enough to determine the great question of a nation's rise or fall, its life or death."[28] China must control its coal or perish.

Now this was hardly a novel argument. The Nanjing School of Mining and Railways had been founded with that very thought in mind—and Liang Qichao had been making similar Darwinesque arguments about the vital importance of this, that, and the other to China's survival in the international struggle for existence for seven years. Nonetheless Lu Xun's Darwinian insistence on the Darwinian importance of coal is interesting and important because it was his first explicit Darwinian argument and because it was a Darwinian argument in two ways at once.

Part of his argument was honestly scientific. Perhaps more clearly than any Chinese before him, he described with Western terminology the succession of geological epochs that had led to the formation of coal, and somewhat gratuitously he described the biological evolution of flora and fauna that accompanied those epochs, summing up his survey with the rather droll comment that nature was an open book, "*Evolution*, written by the creator himself, which Darwin plagiarized in order to become a great nineteenth century author."[29]

But having established that coal and the Chinese race were both products of evolution, he argued that the one would be the *sine qua non* of survival for the other, in its unending evolutionary struggle for existence. Coal was the evolutionary weapon (an odd concept) with which the Chinese could beat back—and this was Lu Xun's term—the "White Peril" (*bai sheng*).[30] Coal was "the hope at the bottom of Pandora's box." For "if we grasp it," wrote Lu Xun, "each day shall see us nearer to a bright and glorious future. If we lose it, all we will be able to do is bewail our inevitable death. Countrymen, we must make the right choice."[31] For any people who lost control of their geological resources would become "fossils, to be picked up and sighed over as relics of an extinct species."[32] The Chinese race would be one with poor Yorick.

This was evolutionary overkill. The Chinese race was in no danger of extinction. The Chinese reproductive system was not going to run out of steam because China ran out of coal. The Chinese had lost political control of coal and country three hundred years before—to the Manchus. But the Chinese were more fruitful than ever. If some "blue-eyed white-faced alien race"[33] were now to seize China from the Manchus, why should the Chinese be in greater danger of extinction than were the Indians, who by all reports continued to multiply like mad, slaves though they were in someone else's sunlit empire.

True enough the White Peril sometimes played for keeps. For some races the threat of biological extinction was no joke. American Indians had cause for alarm. The last of the Mohicans could have echoed Lu Xun's rhetoric without exaggeration. But the Chinese were not in danger of extinction. Any imperialists who tried to drain the Chinese gene pool—through acts of either love or war—would die of exhaustion. Lu Xun's rhetoric made no sense.

Why then did this young lover of science say such things? How could he say such things? How could he say of his people, surrounded by imperialists and yet fighting among themselves, that they would be "eliminated by the cosmic process; they will degenerate day by day becoming apes, then birds, then clams, then algae, and finally inorganic things?"[34] That was not evolutionary argument, it was patriotic bombast.

Small wonder. Lu Xun was an angry and exasperated patriot. But patriots who admire his anti-imperialistic patriotism today must do so with clear heads. For although his anti-imperialistic patriotism is admirable, his evolutionary argument is not. He should have known better. If he did not, we must admit that a budding scientist did not yet understand evolution. If he did, we must admit that in the midst of moral indignation he did not give a fig for true science. In either case we must admit that in such moments Lu Xun did *not* speak with scientific authority. We must be prepared to admit the mystery that his rhyme rang truer than his reason.

At any rate, it is in these ringing if unreasonable rhymes that those who have ears to hear hear revolutionary overtones and those who had ears to hear probably heard revolutionary overtones—in 1903. But then as now, to hear one had to want to hear, because Lu Xun's revolutionary overtones were not very loud.

Lu Xun did at one point verbally pound the table and exclaim, "China belongs to the Chinese!"[35] —And those could be fighting words, when voiced by Sun Yatsen and Co. For then they meant, "We Chinese must seize our country from the Manchus!" But Lu Xun never said that. He spoke only vaguely of Chinese protecting China from the pale faces. He never mentioned the Manchus.

Only in one line did he openly raise the specter of revolution. In a strangely inverted argument, if revolution was his aim, he argued that geological strata could be skipped just as stages of social evolution could be. "Those who talk of the history of the human race," he said, "declare that it is a natural law of political evolution to go from a monarchy to a constitutional monarchy, to a republic. But surely one can find in history instances of sudden switches from a stern monarchy to a republic at one stroke of a bloody sword. Changes in geological strata are also like this."[36]

How in Heaven's name, or evolution's, could "revolutions" in the earth imitate the revolutions of mankind? Was the earth to take lessons from earthlings? Changes in geological strata were not at all like bloody coups, nor were bloody coups in any way like changes in geological strata. The most ardent admirers of Lu Xun should admit that as science this simile was ridiculous.

And yet it echoes Sun Yatsen's equally ridiculous rationale for revolution. For Sun Yatsen did believe that just as there were strata in the earth so there were strata in human history, fixed strata and yet strata that could be skipped, by men who had foresight, will power, and bloody swords. Revolution was the way to *lie deng*, to overleap an evolutionary stratum. Revolution was the way to leap into a republic from a monarchy—without having to slog through the stratum of constitutional monarchy.[37]

Of course, Lu Xun was right when he said that history held examples of bloody swords changing monarchies into republics. What was ridiculous was his insistence, and Sun Yatsen's, that geology be dragged into the argument. For fixed or jumpable strata theories of human evolution were not down to earth theories at all. They came from the stratosphere, off the top of Sun Yatsen's head—and Lu Xun's, and Liang Qichao's and Yan Fu's. All of those gentlemen believed in naturally fixed stages of human evolution. The revolutionaries believed they could leap them. The reformers believed they could not. Lu Xun supported Sun Yatsen's great leap theory, although he stated it backwards, and so he supported revolution—at least in one sentence.

Much has been made of that sentence, that "one can find in history instances of sudden switches from a stern monarchy to a republic at one stroke of a bloody sword." Too much has been made of it. Historians in the People's Republic have quoted it over and over again to prove that Lu Xun, "Great Revolutionary" in the making, knew intuitively, *ab origine*, that "*zao fan you li*," that "It is right to rebel."[38] That was the cosmic truth to which all Marxism-Leninism, the Chairman would say, could be reduced. And so that has been the cosmic truth that historians in the People's Republic have sought to establish above all others. But Lu Xun did not establish it, at least not in 1903. His support of it was backhanded at best. At best he *implied* that it was right to rebel. He never said so outright.

The great opponent of revolution, the great Darwinian, gradualist opponent of Sun Yatsen's great leap theory, Liang Qichao, was a hundred times more revolutionary in his rhetoric than Lu Xun was. Liang Qichao said, "Revolution is an inescapable law in the world of evolution." He said, "The work of revolution (what the Japanese call *geming* and what I call *bian-ge*) is today our only way of saving China." He said, "If our people want to survive, they must begin by forcefully advocating a great revolution, and by carrying out a great revolution." He said, "I have held from the beginning in my essays that political revolution is the only way to save the nation."[39]

True, Liang Qichao always managed by the end of each essay to re-translate revolution into reform. Nevertheless his language was inflammatory. Lu Xun's was not. Lu Xun was cautious, in word and, indeed, in deed. Except on the day he cut off his queue.

That was a revolutionary act. And yet he was slow to act even then. He took his cue to cut his queue from his friend Xu Shouchang. But Xu Shoushang cut his queue as soon as he got to Japan—half a year before Lu Xun cut his, and Lu Xun reached Japan half a year before Xu Shoushang. Why did he wait so long?—Perhaps because he faced a greater risk. The watchdog of his class was the notoriously reactionary—or loyal—Qing official, Yao Wenfu, who did not take barbershop rebellions lightly. He could not, it is true, have cut off Lu Xun's close cropped head, but he could have cut off his stipend, and Lu Xun wanted to study. So slow to act though he was, as the first in his class to cut off his queue, Lu Xun showed courage.[40]

As things turned out he was lucky. The day was saved by the antics of a student four years younger than Lu Xun, but hotter of head, Zou Rong, so soon to publish his pamphlet, *The Revolutionary Army*, so soon to be arrested, and so soon to die in prison. Zou Rong and several friends, in the Spring of 1903, learning that Yao Wenfu was overseeing one of his female charges a bit closely, surprised him in a position so compromising that they were able to cut off *his* queue and send him scurrying, without his tail between his legs, back to China, shorn of his dignity and thus unable to exact revenge.[41]

So was Lu Xun rescued. He had defied the authorities, but he had not had to live up to the brave talk he had inscribed on the back of a photograph taken to immortalize his revolutionary hairdo in living black and white:

> My heart cannot flee the arrows of desire
> As wind and rain darken our ancient land.
> Why look to the Cold Star? He does not understand.
> I shall dedicate my blood to China.[42]

The first and third lines of this famous poem remain so cryptic—no one is sure to what Lu Xun's allusions allude—that they are barely worth translating.[43] But the second and fourth lines are clear enough, especially to those who protest too much "that it is right to rebel." They see in them crystal-clear proof of revolutionary resolve. But it was youthful resolve, brave talk but only talk, clear language, but language we cannot take literally, because Lu Xun did not take it literally. In 1903 he joined in Japan a Zhejiang forerunner of Sun Yatsen's revolutionary party, but he never followed any real revolutionaries into battle.[44] In 1903 he cut off his queue, offered his blood to his country, called on men of Spartan spirit to "throw down their pens and arise," and then put up his pen and went to medical school.

True, years later he would write that part of the "beautiful dream" that led him to medical school was "to be a military doctor in time of war," but he did not describe that war as a revolutionary war, and he never went to war anyway.[45] For after two years of study he quit medical school and went back to writing, giving up on his countrymen's bodies to make war on his countrymen's minds. Lu Xun's wars were all pen wars. He never grasped a real sword. His "daggers and spears" were essays. His bravest act would be to walk unarmed to the funeral of an assassinated fellow pensman.[46] By that time, having waged pen wars against the running dogs of warlords and the murderous Guomindang (*not* exaggerated language), he had indeed risked his blood for his country. Still we must distinguish between militant language and militant action, between pen wars and real wars, between struggles to change men's minds and struggles to cut off men's heads. If Lu Xun thought it right to rebel, we must know what he meant by rebellion.

3

To Change Men's Minds

Lu Xun put up his pen in 1904 and in 1906 took it up again with a new sense of purpose, born of the pain of a Paulinesque conversion—or almost. Anyone who knows anything about Lu Xun knows the story. Told so often (Lu Xun told it twice), it will appear to many, I fear, hardly worth the retelling. But we must take it into account.

Lu Xun's biology teacher in Sendai, finished with pictures of microbes, filled out the hour one day with slides of current events—of the Russo-Japanese war. And suddenly there flashed before Lu Xun's eyes a picture of a Japanese soldier about to behead, on Chinese soil, before a Chinese crowd, a Chinese peasant accused of spying for the Russians. And Lu Xun's classmates all shouted, "Banzai!"[1]

Lu Xun did not. Instead, stunned with anger, disgust, and shame, he decided, seemingly then and there, to give up his study of medicine—because his countrymen looked sound enough in body, as sound as the Japanese. Their illness was in their eyes, in their numbed, unfeeling, compassionless, spiritless, indignationless, gawking stares.

You can still see Lu Xun's crowd, in photographs old and not so old. Such crowds existed. They probably still exist—and not just in China. Let any who doubt it look in Alistair Cooke's *America* at one of our own post-lynching lynch mobs, photographed later than Lu Xun's crowd. Our crowd is even worse, because it does not just gawk, it smiles.[2] But Lu Xun's crowd was bad enough, and it obsessed him. It haunted his stories and his essays for the rest of his life.

So that slide did change his life. "At that time and in that place I changed my mind," he said:[3] "I decided that the study of medicine was not an important thing after all. For a weak and stupid people, however physically healthy and strong, could only serve as material for totally meaningless public executions, or as spectators to such executions. There was no need to think it unfortunate however many of them died of disease."[4]—An un-Hippocratic conclusion for a medical student. But he was even more blunt when he told his shocked friend Xu Shoushang why he was withdrawing from medical school: "I plan to study art and literature.

China's cretins, her wretched cretins, how can medicine ever cure them?"[5] "The most important thing," he said, "is to change their spirit." And the best way to do that was to foment "a literary movement."[6]

This was a dramatic change of mind, but not quite as dramatic as Lu Xun's biographers have made it seem. Lu Xun did not fall blinded off his horse. He did not even stride out of the classroom and medical school in the same hour. In one of his two accounts of his conversion, he said he quit before the school year was done.[7] In the other he said he left at the end of the school year, after telling his favorite teacher a white lie to cheer him up—that he would not waste all he had taught him, because he was going to study biology.[8] His formal education, however, was over, except for some more German lessons, a brief fling at Russian, and a very odd half year's study of ancient Chinese philology with the very odd revolutionary, anarchist, and Buddhist, Zhang Binglin.[9] What he really did was start reading (modern Western literature, mostly from "oppressed countries") and writing (a package of five, long, difficult essays). He did indeed turn to literature to try to wake his people up—but to what?

The above drama has been reduced by convention in the People's Republic to a single four-character formula, repeated, with slight variation, in virtually every essay written on Lu Xun's life or thought. Lu Xun is simply said to have "*qi yi cong wen,*" to have "given up medicine and turned to literature"—which he did. But that formula is misleading, and the use to which it is most often put more misleading still. For it resonates all too easily with that other phrase, that ought to be its opposite: "*tou bi cong rong,*" to "throw away one's pen and join the army." Scholars in the People's Republic almost equate *cong wen* with *cong rong.* They imply that in turning to literature Lu Xun did join the army—the revolutionary army, even though only, in his own famous phrase, as "a soldier in the realm of the spirit."[10]

Their rhetoric is uniformly militant. Lin Fei and Liu Zaifu say that "the change of direction in Lu Xun's life in which he gave up medicine for literature took place almost at the same time that he changed politically from reform to revolution. Thereafter he was filled with an intense longing to follow, using art and literature as a weapon, the revolutionaries, in their battle for a new life for their ancestral land. He truly believed that art and literature could save the nation, could change people's spirit and make it possible for the ancestral land to win for itself a new life."[11]

Zheng Yi says, "In short [the progression] from 'Save the nation with science' to medical reform, to the abandonment of medicine for literature was the inevitable result of the development of Lu Xun's thought through elementary revolutionary practice. If one can say that just before and after he went abroad his lofty aim 'to give his blood for the nation' was still only a vague expression of the pursuit, for his ancestral land, of wealth and power, and that he still had no clear idea of what 'the art of achieving wealth and power' was, then one can say that after a four year search through preliminary attempts at revolutionary practice his path had become clear: He would be a 'soldier of the spirit' and taking up the weapon of art and literature hurl himself into the struggle in the realm of ideological consciousness, to call his people to their senses and 'change their spirit.'"[12]

And Wang Shiqing, one of Lu Xun's earliest biographers, uses martial idiom over and over again, as he has since the beginning:

In the summer of 1906 Lu Xun returned from Sendai to Tokyo and thereupon commenced his activities in revolutionary literature.[13]

At the time, the main weapon he used to awaken the people was that of art and literature.[14]

Starting from a militant, materialistic philosophical viewpoint, Lu Xun held that political and intellectual struggle were unavoidable, that you could not avoid them even if you wanted to.[15]

Although there are many facets to the functions of art and literature, the main task is to inspire the people to wake up and dare wage a struggle of resistance against their oppressors.[16]

[Lu Xun] was one of the first 'spiritual warriors' from among the Chinese people to wake up and oppose imperialist aggression and oppression.[17]

But was a "spiritual warrior" a warrior? This is no "Is-a-white-horse-a-horse" kind of question.[18] If a real warrior was one who really made war—who was willing to kill—or who urged others to make war, we must confront all this militant rhetoric and put it to the test: Did Lu Xun or did he not in 1906 become a real warrior?—The question is *not* easily answered.

Militant rhetoric can, of course, be rhetoric only. The militant rhetoric of Christian missionaries in China in Lu Xun's day is proof enough of that. John Hersey's 1985 novel, *The Call*, a remarkable work of fiction *and* history, positively bristles with military metaphors, true echoes of the language of true missionaries.

Missionaries at home rallied recruits with language like this:

This is a council of war. In the tent of the Commander we are gathered, and the Commander-in-chief is here. We are his subordinates, the heads of departments, the under captains, and here are the volunteers of the army.[19]

In Beijing, members of the World's Student Christian Federation plotted "the Christian Conquest of China."[20] In the field, "Christian soldiers" steeled themselves for "battle":

It is war, and I as a general must be ready to face even death.[21]

But when "Christian soldiers" did face death, some showed themselves, at the last, not soldiers at all:

How I honor the compassion in the face of death of Dr. Taylor, who showed the Boxers his pistol and said he disdained to take human lives and threw the pistol into the flames of the Lowrie house, in which flames he then died.[22]

Scientists too could be misleadingly militant, evolutionists more easily than any. The discovery of evolution, said Huxley, had given rise, among other things, to "the gladiatorial theory of existence," which he repudiated as a way of life but did not deny as a description of all too much of it.[23] He himself had said that "the animal world is on about the same level as a gladiator's show." He himself had said that at least for primitive man, "Life was a continuous free fight, and beyond the limited and temporary relations of the family, the Hobbesian war of each against all was the normal state of existence."[24]

Railing against such rhetoric, in *Mutual Aid*, Peter Kropotkin sought to remind us over and over again of Darwin's own warnings not to take the most militant rhetoric of *The Origin* too literally:

I should premise that I use the term Struggle for Existence in a large and metaphorical sense. . . .[25]

Wielding that sentence over and over again, Kropotkin attacked other fighting words in *The Origin:*

The same remark which Darwin made concerning his expression: "the struggle for existence," evidently applies to the word "extermination" as well. It can by no means be understood in its direct sense, but must be taken "in its metaphoric [*sic*] sense."

"'Extermination,'" he added, "does not mean real extermination," and "what is described as competition may be no competition at all."[26]

Happily we need not resolve, at this point, this real Darwinian struggle over the meaning of Darwinian struggle.[27] It is enough, perhaps, to say that Huxley saw too much warfare in the world and Kropotkin too little. But we must note one thing: Kropotkin himself was not above occasional flourishes of military metaphor. "Life in societies," he said, "is the most powerful weapon in the struggle for life."[28] Metaphorically or not, Kropotkin got himself into semantic difficulties, even as he tried to get us out of them, just as Lu Xun got himself into semantic difficulties that his communist countrymen have kept us in.

Was Lu Xun's writing really "a weapon," his pen really a sword? Was he really "a warrior?" Or were his "revolutionary activities" not revolutionary activities at all? When Chinese scholars echo in essay after essay Chairman Mao's words that Lu Xun was "a great revolutionary," should the word revolutionary "by no means be understood in its direct sense?" Should it "be taken in its metaphoric sense" only? Or should it be taken for real?

In the immortal words of Mencius, "It is hard to say."[29] For what is "for real?" For me "real revolution" is the revolution of the French Revolution not the Copernican Revolution. Meaningful as the word revolution is in its "metaphorical sense," I think we should put that metaphorical sense aside and judge the Chairman's judgment of Lu Xun with the Chairman's own definition of revolution: "Revolution is no dinner party— Revolution is violence."[30] For me, "real revolution" is struggle for change that countenances killing, and the question we must ask, and, far more importantly, that Chinese must ask, if in the future they seek advice from Lu Xun, is, did he, when he gave up medicine to take up "the weapon of literature," countenance killing?

"It is hard to say." Chinese scholars can still not agree on whether or not Lu Xun really joined a revolutionary party when he returned to Tokyo—or for that matter on whether or not he had already joined one before he left Tokyo for Sendai. Most Chinese biographies or chronologies of his life dutifully state that although he was never a member of Sun Yat-sen's Tong Meng Hui (the Alliance), founded in 1905, he was a member of the Guangfu Hui (the Restoration Society), founded in 1904. As "a great revolutionary," of course, he should have been. But Ni Moyan, author of the most exhaustive and objective study of the problem that I have seen, admits that there is as yet no hard evidence to settle the matter—no membership list or the like. Nor are there any statements from other party members saying that they worked with Lu Xun as a party member. The soft evidence is most puzzlingly contradictory. Lu Xun's relatives suggest that he did not join the Restoration Society. His best friends protest that he did.

His first younger brother, Zhou Zuoren, who was with him in Tokyo in 1906, said, "He never did join the Alliance, although he spent a lot of time in the offices of the *Min bao* (the People's Journal, the "Real Revolutionaries" most important publication) and although many of the people he had contact with were Alliance people. He did not join the Restoration Society either. —Why not? I don't know." His youngest brother, Zhou Jianren, agreed: "Lu Xun's relations with Wang Jinfa [head of the Restoration Society] were very close. He was in touch with all of the leaders of the Restoration Society, and he knew all about the Restoration Society, but he never told me he joined it." Of course maybe he was not supposed to tell. The Restoration Society was, as Ni Moyan reminds us, a secret society. But why, years after the revolution, should he have kept his membership that much of a secret? His eventual wife, Xu Guangping, for all she said about it, seemed in the dark to the very end. Lu Xun knew all about the Restoration Society, she later said, when asked whether he had been a member, "but he was apparently not yet willing to take part in any practical action."[31] Did that mean that Lu Xun was not a member or only that he was not an "active" member?

Lu Xun's best friend, Xu Shoushang, later claimed that "Jingsong [Xu Guangping] knew perfectly well that Lu Xun was a member of the Restoration Society." But why would she not say so? Xu Shoushang himself is said by all to have been a member of the Restoration Society since its founding in 1904. But he said that Lu Xun did not join when he did. He later said that Lu Xun joined in 1908, but that he did not learn that Lu Xun had joined until 1909, when they both returned to China. But from 1905 to 1909 the Restoration Society had lost itself in the Alliance, which Lu Xun said he did not join! If Lu Xun was indeed a member, he must have been one of the most secretive members the Restoration Society ever had.[32]

Nevertheless, Lu Xun's best friends in the thirties, Feng Xuefeng and Hu Feng, both said Lu Xun told them he was a member—sort of. Feng Xuefeng said, in one line, that Lu Xun told him he "belonged to the Restoration Society," but Ni Moyan has conscientiously pointed out that that Chinese phrase could just mean that he was "pro" the Restoration Society or "on its side."[33] Hu Feng said Lu Xun told him, "I joined the Restoration Society, but nobody knows about that."[34]—and still nobody knows. Shen Diemin, an early Tokyo roommate of Lu Xun, who later unquestionably did become a real revolutionary, matter of factly said that both Lu Xun and Xu Shoushang joined the Restoration Society the year it was founded, in 1904.[35] But again—it was swallowed up by the Alliance from 1905 to 1909, and Lu Xun said he was not a member of that. So how active a real revolutionary could he have been in those important years in Tokyo after he abandoned medicine for "revolutionary literary activity?" And how active a member of the Restoration Society could he have been in 1904—all alone in Sendai?

Still there is one more piece of evidence that suggests he was a member. To Ni Moyan it is the piece that ultimately proves it. To me it does not, because there is still something odd about it. But perhaps, nonetheless, it proves something more important.

From April to July 1931, a Japanese scholar of Chinese literature, Masuda Sho sought Lu Xun's help with a translation he was working on of Lu Xun's *Zhongguo xiaoshuo shilue* (a brief history of Chinese fiction). He also wrote a biography of Lu Xun and showed Lu Xun his draft—in which he said Lu Xun had been a revolutionary party member. Years later, Mr. Masuda argued that Lu Xun must have been a party member, because when he read the draft he did not deny it. But he had to *infer* that he was a member. He did not claim that Lu Xun told him he was.[36] He had to infer that Lu Xun was a revolutionary party member even from this story, which Lu Xun did tell him:

Lu Xun told me: "When I was engaged in anti-Qing revolutionary activities, I was once ordered to go assassinate someone [this was deleted from the draft of my *Biography of Lu Xun*]. But I said, 'I could go, and I might die, but dead I would abandon my mother. I ask you what should I do about my mother?' And they said, 'It's no good if you're worried about things after you're dead. You needn't go.'"[37]

Masuda Sho argued, to his own satisfaction, that only a revolutionary party member could be *ordered* to commit an assassination, but if his story is true, the question of Lu Xun's party membership loses its importance.[38] What counts is that Lu Xun was given a chance to engage in revolutionary violence, and he did not leap at it. He had a chance to kill for the cause, and he did not take it.

Why not? We cannot be sure. Maybe he was a filial son. Maybe he did not quite have the courage. Maybe he thought it was not yet the moment. Maybe he thought assassination would not accomplish anything—so that killing and dying were not yet worth it. That is what he seemed to think of the revolutionary violence planned by Xu Xilin and Qiu Jin, and of its tragic aftermath. Lu Xun had argued against Qiu Jin when she had tried to get all the Chinese students in Japan to go home in united protest against the Japanese government's attempts in 1905 to curtail politically embarrassing Chinese student anti-Qing agitation. Sounding like Hu Shi in 1915 arguing with Chinese students in America who wanted everyone to go back to China to protest Japan's infamous "Twenty-one demands," Lu Xun said that a patriotic student's first duty was to study. But Xu Xilin and Qiu Jin went back. Qiu Jin started a school, provocatively offering her students military training. Xu

Xilin assassinated the Manchu governor of Anhui on July 6, 1907. Captured within hours, he was executed the next day, and had his heart cut out. Qiu Jin was captured, tortured, and beheaded a week and a half later.[39] Lu Xun protested their executions, but he also said Qiu Jin was "clapped to death."[40] He seemed to believe that the applause she had won with her revolutionary rhetoric in Japan had gone to her head, and so she had lost it—vainly.

But to say that Qiu Jin was clapped to death was not to say that assassination was bad. To hesitate out of filial piety was not to say assassination was bad. Not to go himself was not to say that others should not go. Perhaps Lu Xun had merely appraised himself and decided that he could do more good writing essays than throwing bombs—probably a good appraisal. Chairman Mao would later say that even charcoal burners served the cause,[41] but it is hard to imagine that Lu Xun could ever have been a good shot *or* a good charcoal burner.

So what can we say? We can accept William Lyell's assessment that "during his years in Japan Lu Xun was no political activist."[42] More specifically, we can accept Ni Moyan's conclusion that Lu Xun "was not enthusiastic about assassinations or bombings." He "valued intellectual struggle" and "entered the war with his pen."[43] And so, refusing to take military or revolutionary rhetoric literally, we can accept Wang Shiqing's statement that "the revolutionary road [Lu Xun] chose was completely different from Xu Xilin's. Lu Xun did not advocate assassination. His idea was to wage a long, protracted, yielding and yet unyielding war of advance and retreat, a war of attrition, to transform the soul of a mute people, to strike sparks in the spirit of a mute people—through art and literature."[44]

But if that was his way, should we not also accept the conclusion of Li Zehou, the only Chinese scholar audacious enough to state that Lu Xun in 1907 and 1908, after abandoning medicine for literature, "differentiated himself day by day from the advanced people of the time and from the vast majority of bourgeois revolutionaries?"[45]

But *how* did he differentiate himself? Was he more revolutionary or less? He did not choose to be an assassin, but did he never applaud assassins? He did not choose to kill, but did he never urge others to kill? He did not choose to be violent, but did he never exhort his countrymen to violence? We must ask these questions of the first essays he wrote after choosing to be a writer. And we must ask if he could daily differentiate himself from real revolutionaries—and be a real revolutionary.

4

Deaf Ears

In 1906 Lu Xun dramatically "abandoned medicine and turned to literature." But no one noticed—aside from a handful of friends and relatives. He, his brother, Zhou Zuoren, and his best friend, Xu Shoushang, struggled to produce a journal—the obvious thing to do, he later said. But they struggled in vain.[1] They meant to call their journal *Xin sheng* (New Life), but behind their backs not so friendly friends said *xin sheng* meant "new students," or "the freshmen."[2] At any rate their "New Life" was never born. Backers backed out, and it aborted.

Lu Xun then sent manuscripts to Shanghai, to the Commericial Press. But they came back unopened, the last with a note: Send no more of these, please.[3]

He wrote on anyway. And finally the founders of another fledgling journal, *Henan,* a journal fitter than *Xin sheng*, fit enough at least for ten issues, though it found no lasting niche, asked him to contribute something, even though Henan was not his province.[4] They needed manuscripts. He had some, and was willing to write more, and so finally, over the course of a year, from December 1907 to December 1908, he published a package of five articles—or four and a half. The last was to be continued. But it never was, because after that half, the whole effort of which it was a part was discontinued—for ten years.

So those four and a half articles form a package apart, a package separated from Lu Xun's first articles by his medical study, by a modest growth in maturity, a clearer sense of purpose, and an even greater obscurity of style, a package that was his second attempt to speak to his people, but his first attempt after consciously dedicating himself to the cause of laboring for the salvation of his country in the vineyard of literature, a package that was his first clear failure—no, as things would turn out, his only clear failure, a failure at any rate so clear to Lu Xun, that it led him into ten years' silence.

That silence itself has set his four-and-a-half-piece package apart, so far apart that intellectual historians have naturally seen in that package "the culmination of Lu Xun's early thought." But the true culmination of

Lu Xun's early thought is in the ten years' silence, as much as it is in the four-and-a-half-piece package. It is in the tension between the two, in the "contradiction" between the two, in the irony of the silence that followed Lu Xun's first conscious attempt to combat silence.

Twenty years later, in 1927, Lu Xun would give a famous lecture (in the Hong Kong YMCA, of all places) entitled "Wu sheng de Zhongguo" (Silent China).[5] Ten years after he broke his own ten years' silence, he would still lament the silence he first lamented twenty years before. The provocation of silence, the provocation to silence—that was not just a contradiction of his early thought. It was a contradiction that would last, that would remain unresolved even at the end, even though after 1917 he would never again keep absolutely silent.

In 1907 it was China's silence that made him "abandon medicine and turn to literature." The key phrases of his speech were already in his head: China, threatened from within and without, was still *"jimo er wu sheng,"* silent, like a place deserted, a place from which came "no sound."[6]

But what did he mean? Were Chinese intellectuals in the first decade of the twentieth century not more raucous than silent? Was the first decade of the twentieth century not the decade of the great (Darwinian) debate over reform or revolution? Liang Qichao in his *Xin min cong bao* (The new people's journal) poured forth impassioned essays for reform, that breathed revolution.[7] Sun Yatsen and Co., in angry, if less articulate, response in the *Min bao* (The people's journal), cried out directly for revolution, and for a Han-ruled socialist republic. And Wu Zhihui and Li Shizeng, and other anarchists, in their *Xin shiji* (The new century), added uncoordinated anti-Manchu antimonarchy shouts from Paris that even if vulgar were at least always loud. Why did Lu Xun hear no sound?

He heard no sound because all the noise came from intellectuals in exile, or from intellectuals writing from the semisafety of the treaty ports. Lu Xun was listening for a response from the Manchu-Chinese government or from the people, from the gawking crowd in that slide. And he heard none.

Throughout his entire life, noise from other intellectuals never seemed to break his sense of silence. He heard few sounds to echo. Before 1911 he applauded neither reformers nor revolutionaries. After 1911 he did not applaud the revolution. He did not applaud the Second Revolution. He briefly echoed Cai Yuanpei's call for aesthetic education, a call heard by almost no one.[8] He wrote one veiled paean of very faint praise for Sun Yatsen, after Sun Yatsen was dead. He praised Hu Shi, once, for advocating the "Literature Revolution," but more than once made fun of him.[9] He praised students killed in demonstrations, but rarely echoed calls to demonstrate. He praised young writers who were executed. He praised a few young writers who were not executed. But he rarely praised the work of authors who succeeded. His peers did not impress him. He had few comrades. Only near the end did he claim to hear distant thunder.[10] Only near the end did he say he would be proud to be called a comrade of those actually fighting for the survival of the Chinese people.[11] But how close a comrade was he? That is still a question. But I think even at the end Lu Xun was lonely.[12]

Clearly, when he gave up writing in 1908, after publishing his four-and-a-half-piece package, he felt the despair of an unheeded prophet in the wilderness. He had said that China's salvation depended on true individuals, "spiritual warriors," who could defy the crowd.[13] And clearly he had hoped to be one, although he disclaimed any messianic ambitions: "Most things are lost through pride, but still there is more hope in launching a reed boat than in waiting for someone else to build an ark. I have not yet given up my great hopes for the future. That is why I have written this."[14]

But when no one listened, he said:

> I felt a senselessness and futility that I had never experienced before. . . . At first I did not know its cause. But later I thought, if you advocate something and people approve, you are encouraged to go forward. If people disapprove, you are encouraged to struggle. It is only when you cry aloud in the midst of the people and get no response at all, neither approval nor disapproval, as if you were all alone in an endless wasteland and there was nothing you could do—that is what is sad. And so I knew I felt loneliness.
>
> And this loneliness grew, day by day, like a great poisonous snake. Coiling around my soul.
>
> And yet, although I felt boundless grief, I was not resentful, because this experience made me examine myself, and see myself: I was definitely not a hero who could gather crowds of followers with a shout and a raised fist.[15]

He decided he was not to be a charismatic leader. Did that mean that he had hoped to be one? Anyway, he recognized that his writing had failed. And it had. He had cried aloud in the wilderness and his people had paid him no heed. Why?

Because neither his message nor his medium was fit.

His medium could not have been less fit. Having espoused the cause of "saving the nation through literature," he espoused a literary style that verged on the suicidal. He wrote like a literary warrior with a literary death wish—in language that could never have raised a following because it was simply too hard to understand. He mired his message in a morass of ancient idioms, archaic characters, and obscure, if not obscurantist, allusions—through which he tried to introduce the thought of a host of unheard of Westerners. I am sure that very few readers of *Henan* bothered to wade all the way through his articles. And probably not many people read *Henan* anyway.

Granted, Chinese readers in 1907 were much better prepared to read Lu Xun's early articles than Chinese readers today. Today Chinese scholars have felt it necessary to provide potential readers—Chinese readers, not just ill-educated foreign readers—not only with pages of explanatory footnotes, but with full vernacular translations! Wang Shiqing has helped make four of the four and half essays in question more accessible, by adding to 96 pages of original text 47 pages of footnotes and 92 pages of translation, both of the latter printed in smaller type. And Zhao Ruihong has gone to even more trouble. He has devoted a 304-page book to a single essay, explaining a 47-page text in 523 footnotes (131 pages in smaller type)—and even then adding a translation! Today, Lu Xun's *Henan* essays are not easily read.

Nor were they when Lu Xun wrote them. When he decided to reprint them twenty years later, he confessed that "influenced by the *Min bao*," he had then "liked making up queer sentences and writing archaic characters," some of which he now agreed to change, "to make the typesetting easier." Indeed he confessed: "Were such difficult things anyone else's, I am afraid I could not help suggesting that they be given up." Why, then, did he not give them up? "One small reason" was that "no one has ever again mentioned the poets spoken of in them."[16]

That confession proves two things. It proves that he had indeed failed. He and his poets had inspired no one (in volumes of memoirs of Lu Xun, I have found no one who remembers being inspired by Lu Xun's early articles). But it also proves that twenty years after his failure, he still believed in his message.

Why he chose to make it so unclear is unclear. Perhaps he simply could not resist showing off his classical learning, which was, indeed, prodigious. But he must too have really liked his "strange sentences." He had not yet rebelled, even in his own head, against the aesthetic that he would later ridicule so devastatingly (which he could do only because he knew so well what he ridiculed). One of his models was still Yan Fu.[17] Another, far less fortunately, was Zhang Binglin, whose writing even Lu Xun had trouble reading.[18] (He might later claim to have studied with Zhang Binglin "not because he was a scholar, but because he was a learned revolutionary," but what he studied with him was classical Chinese phonology and etymology, not revolution.[19]) At any rate, although he had given up his Spartan bombast, many passages in his second package were stilted and pedantic. And yet, other passages did have an old-fashioned power, if one could understand them. Zhao Ruihong protests even now that "many passages read aloud very well and are quite rich in musical beauty."[20] But as Song Yu pointed out twenty-three hundred years ago: "*Qu gao he gua*" (when the piece is high few can harmonize). Lu Xun's music did not inspire masses even of intellectuals to "join in rebel chorus"[21] (the crowds that he sought to wake up, of course, could not read at all). Neither music nor message had a prayer of proving popular.

But what, in his first vain attempt to help "save the nation through literature," had he said?

Here is the package:

1. "Ren zhi lishi" (The history of man)
2. "Kexue shi jiao pian" (Lessons from the history of science)
3. "Wenhua pian zhi lun" (On the zigzag progress of civilization)
4. "Moluo shi li shuo" (On the power of Satanic poetry)
5. "Po e sheng lun" (On dispelling false ideas)

And here, in brutal extraction, is its message:

1. Progressive evolution is a proven fact. We Chinese must accept it. Animal ancestry is no disgrace. Man is still the highest being. Religious superstition has been bowled over. All is clear. Living matter came from non-living matter. All that remains is for cosmogony to explain where non-living matter came from.[22]

2. The remarkable progress of the modern world owes most of its existence to science. Religion, more often than not, has resisted science, so the way of progress has not been straight, but twisted like a spiral. But this has not been altogether bad, for religion too has made its contributions, especially to morality and art, and to the moral force and idealism that have inspired scientific endeavor itself. So we cannot just pin our hopes on things. We need knowledge and idealism—science, art, and literature.[23]

3. For millennia China has been the fittest in the East. But we have had no real competition. Now we must struggle to survive. Our survival depends on people of spirit not on things. We need true individuals, who dare defy the crowd. We must beware of the crowd. We must beware of the tyranny of the mindless masses, as well as the tyranny of greedy people who would rule in the name of the masses. Progress is made by those who resist the status quo, but in turning against it,

people often go too far. We must resist our past and present, but try not to go to extremes. But to accomplish anything we must have the right people. So we must respect individuality and engender spirit (*zun gexing er zhang jingshen*).[24]

4. We must learn from Byron, and other "satanic" poets. We must fight the status quo. Our ancient civilization enfeebles us. Progress is for those who pursue perfection. We must struggle to rise above the stupid crowd. We must resist imperialist aggression—but not become imperialists ourselves. We must give men of genius room. We need free warriors of the spirit. We need men of vision to teach us new culture.[25]

5. China is doing itself in—and still no one protests! We need individuals who will speak their minds, even if all are against them. We need a nation of individuals. But our mean spirited masses now put down men of talent. And so China remains silent. Or if these are new voices, they do us no good. People go to Europe and bring us back the art of corseting, proclaiming the uncorseted uncivilized! Or they preach nationalism or internationalism, *both* of which destroy individuality. Or they say that religion is superstition and claim that science explains everything. But we need to believe in *something*. Some say we just need patriotism, like the patriotism of those who threaten us. But theirs is brutish patriotism, the inheritance of their brutish ancestors. They are not yet fully evolved. We must defend ourselves, but we must act like human beings. We must defend ourselves, and if we have any extra strength, help other victims of aggression.[26]

This was not a remarkable message. Indeed, except for his farewell hint of a call for "third world solidarity" (rarely, heard in those days[27]) and his idiosyncratic defense of individualism, Lu Xun's message was more remarkable for what it did not include than for what it did. Although he praised his "Satanic poets" for their spirit of resistance, the "great revolutionary" still spoke clearly only of resistance to foreign aggression. He breathed not a syllable about resistance to the Manchus and only two syllables about resistance (an "alas") to monarchy: "*Wuhu*," he said, "In the past people were ruled by single despots. Now suddenly there has been a change: The modern way is to be ruled by ten million good-for-nothings"[28]—not a change for the better: "Suppression in the name of majority rule is even worse than suppression at the hands of a tyrant."[29]

But if he disliked democracy as much as monarchy, what was he for? He was for the oldest Confucian cure-all in the book: *li ren*—"give power to the proper people."[30] But who was to give power to whom, and how? Lu Xun gave no answer. The proper people were true "individuals." Maybe they were Nietzschean supermen.[31] But where were they? Lu Xun pointed to no flesh and blood *Übermenschen* on the horizon. Most pointedly he did not point to Sun Yatsen and Co. Instead he looked for people who would "introduce new culture."[32] That made "the proper people" look more like Lu Xun himself than like Sun Yatsen—and much more like reformers than revolutionaries. Lu Xun should not have been surprised that he went unhailed as a writer of revolutionary charisma.

But why, then, should we now pay attention to these essays, if no one paid attention to them then? —Because, despite all, we can see in them, in the culmination of Lu Xun's early thought, the culmination of Lu Xun's early understanding of evolution. And in that, we can see convictions and contradictions that would affect his thought for the duration.

5

The Riddle of the Universe

Lu Xun's "History of Man" proclaimed, pretty much, the gospel according to Ernst Haeckel.

There is no understanding Lu Xun's "history" without understanding Haeckel's "gospel." So we must take time here for what may seem a digression.

It is *not* a digression.

Ernst Haeckel was Darwin's doberman, if Huxley was "Darwin's bulldog."[1] Professor forever at the University of Jena, billed by one of his English translators as "one of the most prominent zoologists of the century" (the nineteenth, of course), and as "one of the leading combatants" "from the scientific side" in "the most salient feature of the nineteenth century—the conflict of theology with philosophy and science,"[2] Haeckel was Darwin's fiercest defender in Germany, Darwin's fiercest defender on the continent, and through the many immediate translations of his many works, one of Darwin's fiercest defenders round the world. In volume after repetitious volume he indefatigably proclaimed the truth of evolution and propagated "Monism," the only fit philosophy, to his mind, for those in the evolutionary know.

Round the world, in their day, Haeckel's works had an untold (and untoward) influence—that reached Japan, and there, Lu Xun, and, through Lu Xun, China. Haeckel was Lu Xun's second mentor, after Huxley, in what Haeckel even more than Huxley led Lu Xun to believe was the cut-and-dried science of evolution, the science of good news that Lu Xun sought to give his countrymen in his very first effort to "save the nation through literature."

In the subtitle to that first effort, Lu Xun promised "An Exposition of the German Herr Haeckel's Monist Studies of Phylogeny,"[3] but just which of Haeckel's studies he had at hand is not clear. The editors of Lu Xun's complete works assumed both in 1964 and 1981 that he was mainly introducing Haeckel's *Anthropogenie* (The Evolution of Man, 1874).[4] Lin Fei and Liu Zaifu imply he was introducing *Die Welträtsel* (The Riddle of the Universe, 1899), having found that there already existed at the time a Japanese translation.[5] But Lu Xun must have known of both works. He mentioned the former and quoted from the latter, even giving a phrase and a term or two in German. So he may have read *The Riddle of the Universe* both in translation and in the original. There is evidence that he knew also of *The History of Creation*.[6] I cannot say that he knew of *Monism*, but we must look at all those works, to understand the whole of Haeckel's thinking.

Joseph McCabe, translator, promised his readers they would "soon discover" in *The Riddle of the Universe*, "a vein of exceptionally interesting thought."—And Lu Xun did. No matter that today Haeckel's works are rarely read. No matter that they have fittingly followed the way of the unfit. No matter that Haeckel himself is remembered, when he is remembered, only as the author of an error—or as a racist. To understand Lu Xun's thought, we must look long and hard at Haeckel's.

Ernst Haeckel, "in matters vegetable, animal, and mineral" was the very model of a nineteenth-century know-it-all. In 1877 he could see "the truth" in a single cell:

> The cell consists of matter called protoplasm, composed chiefly of carbon, with an admixture of hydrogen, nitrogen and sulphur. These component parts, properly united, produce the soul and body of the animated world, and suitably nursed become man. With this single argument the mystery of the universe is explained, the Deity annulled and a new era of infinite knowledge ushered in.[7]

As a nineteenth-century know-it-all, his only rival for best of breed was Herbert Spencer.

To be fair, Haeckel did not himself claim infinite knowledge. "At the close of the nineteenth century," in 1899, in the preface to *The Riddle of the Universe*, he admitted that "the studies of these 'world-riddles' which I offer in the present work cannot reasonably claim to give a perfect solution to them."[8] Nonetheless in his conclusion he dared declare that "only one comprehensive riddle of the universe now remains—the problem of substance," which Haeckel did not think a substantial problem:

> We grant at once that the innermost character of nature is just as little understood by us as it was by Anaximander and Empedocles. . . .We must even grant that this essence of substance becomes more mysterious and enigmatic the deeper we penetrate into the knowledge of its attributes, matter and energy, and the more thoroughly we study its countless phenomenal forms and their evolution. We do not know the 'thing in itself' that lies behind these

knowable phenomena. But why trouble about this enigmatic 'thing in itself' when we have no means of investigating it, when we do not even clearly know whether it exists or not? Let us, then, leave the fruitless brooding over this ideal phantom to the 'pure metaphysician,' and let us instead, as 'real physicists,' rejoice in the immense progress which has been actually made by our monistic philosophy of nature.[9]

The place of the "unknowable" was for Haeckel as for Spencer a place of no account.

So he rejoiced in nineteenth century knowledge, in a manner proud enough to make a peacock molt. At the close of the nineteenth century he looked to the future with breathtaking confidence—that somehow came from the theory of evolution:

Our Theory of Development explains the origin of man and the course of his historical development in the only natural manner. We see in his gradually ascensive development out of the lower vertebrata, the greatest triumph of humanity over the whole of the rest of Nature. We are proud of having so immensely out-stripped our lower animal ancestors, and derive from it the consoling assurance that in future also, mankind, as a whole, will follow the glorious career of progressive development, and attain a still higher degree of mental perfection. When viewed in this light, the Theory of Descent as applied to man opens up the most encouraging prospects for the future, and frees us from all those anxious fears which have been the scarecrows of our opponents.[10]

"Our opponents," "our theory"—that "our" meant both "of us the nineteenth century cognoscenti" and "of Darwin and me (and a few selected others)." But although consciously imitating the *Origin* at the end of his *History of Creation*, Haeckel, in his confidence, left Darwin in the dust.

Cautious Darwin said, "We can dimly foresee that there will be a considerable revolution in natural history."[11] And he said, "Light will be thrown on the origin of man and his history."[12] And then, in his most unguarded moment, he said what, "in theory," he had no right to say: "As natural selection works solely by and for the good of each being, all corporeal and mental endowments will tend to progress towards perfection."[13]

He did not attempt to say what he meant. (It is probably just as well), but to Haeckel, as usual, all was clear:

We can even now foresee with certainty that the complete victory of our Theory of Development will bear immensely rich fruits—fruits which have no equal in the whole history of the civilization of mankind. Its first and most direct result—the complete reform of *Biology*—will necessarily be followed by a still more important and fruitful reform of *Anthropology*. From this new theory of man there will be developed a new *philosophy*, not like most of those airy systems of metaphysical speculation hitherto prevalent, but one founded upon the solid ground of comparative Zoology. A beginning of this has already been made by the great English philosopher Herbert Spencer. Just as this new monistic philosophy first opens up to us a true understanding of the real universe, so its application to practical human life must open up a new road towards moral perfection.[14]

Again to be fair, we must admit that however proud Haeckel was of the "marvelous progress"[15] of the nineteenth-century, he had no illusions as to the nineteenth-century moral state even of "the English and Germans," "chief representatives"[16] of "the *Caucasian, or Mediterranean* man (Homo Mediterraneus)" who "has from time immemorial been placed at the head of all races of men as the most highly developed and perfect."[17] Several times over in his works he echoed the sober self-assessment of Alfred Wallace:

Compared with our astounding progress in physical science and its practical application, our system of government, of administrative justice, and of national education, and our entire social and moral organization remain in a state of barbarism.[18]

So "moral perfection" belonged to the future. But still Haeckel's optimism is breathtaking. For he honestly believed that we were on our way to perfection, that evolution was about to lift us forever out of barbarism. He even dared "express the hope that the approaching twentieth century will complete the task of resolving the antitheses" of "theism and pantheism, vitalism and mechanism,"[19] which were still the major obstacles to "the victory of the natural idea of 'Man's Place in Nature,'" through which "the progressive development of the human spirit will be advanced in [such] an unusual degree"[20] that verily an *existence worthy of man, which has been talked of for thousands of years, will at length become a reality.*"[21]

The nineteenth century, the Age of Optimism, the great age of secular superstition—as silly as the other kind. Haeckel's breed has not died out, but at the close, or almost the close, of the twentieth century, purebred specimens no longer abound. Looking back on the most sanguinary century of all time, who can be as sanguine as Haeckel about the future? We see in the mirror of our history a creature who threatens to make lemmings look bright, a creature we know is clever enough and have good reason to fear may indeed prove stupid enough to commit sui-genocide or geni-suicide,[22] to outdodo the dodo—which was not responsible for its demise.

Poor Haeckel lived until 1919. "Wholly [and so proudly] a child of the nineteenth century,"[23] he might more happily have died in 1899, unless like Condorcet he was able to keep the faith to the end, even as his world went up in smoke. But whether he kept it or lost it, to evaluate his influence on Lu Xun, we must ask where Haeckel got his faith and whether or not he had indeed those "good grounds for belief"[24] that his far less optimistic, agnostic colleague, Huxley, insisted we must always have before we believe in anything. *Nota bene:* It was the agnostic Huxley, who could not believe in "modern speculative optimism with its perfectibility of the species, reign of peace, and lion and lamb transformation scenes,"[25] who was the *rara avis* in the nineteenth century, not Haeckel. Optimism was common. But where did Haeckel's come from?

It came, Haeckel said, from his study of "the course of the History of Evolution."[26] To Haeckel's own satisfaction, evolution led inevitably to "Monism," and Monism inevitably to optimism. To each his own logic. I confess I can only see evolution leading explicably but evitably to Monism and Monism leading inexplicably, and thus even more evitably, to optimism. To my mind, Haeckel ultimately reached optimism by a leap of faith that owed more to the "faith of our fathers," which he found "an indefensible superstition,"[27] than to the "palace of reason" he thought "modern science" had built on "the ruins" of that superstition.[28]

In a moment of unrecognized veracity, Haeckel subtitled an apology for Monism "A Confession of Faith of a Man of Science,"[29] and that is indeed what it was. But within it he confessed no confession of faith at all. All was "knowledge," based on the Darwinian revelation of man's "true 'place in nature':"[30] "Man is not above nature, but in nature."[31] Huxley's "'question of all questions [sic]'" was "scientifically answered: 'Man is descended from a series of ape-like mammals'"[32] and "secondarily from a long series of lower vertebrates," and they from a "chain of our earlier invertebrate ancestors" all the way down to "a unicellular ancestor, a primitive Laurentian protozoon,"[33] and indeed even beyond, to a "little lump of protoplasm" that "did not yet possess a cell-kernel," to a "monera" that "*originated* in the beginning of the Laurentian period by *spontaneous generation, or archigony,* out of so-called 'inorganic combinations' of carbon, oxygen, hydrogen, and nitrogen,"[34] which "elements" themselves were nought but—"substance." That is all there was. There wasn't any more.

Monism was built on "the law of substance," "the supreme and all-pervading law of nature, the true and only cosmological law,"[35] that proclaimed "the essential unity of inorganic and organic nature, the latter having been evolved from the former."[36] "All the wonderful [strange word] phenomena of nature around us," said Haeckel, "organic as well as inorganic, are only various products of one and the same original force, various combinations of one and the same primitive matter."[37] Combinations come and go, "yet in this 'perpetual motion' the infinite substance of the universe, the sum total of its matter and energy, remains eternally unchanged," for "over all rules the law of substance."[38]

One substance, one law, one evolution, one philosophy: Monism. But dualism crept in from the start. Were substance and the Law of Substance, nature and natural law, one or two (an ancient problem)? Also, Haeckel's universe had no space: "Mass," "ponderable matter" was suspended in a sea of "ether," "imponderable matter."[39] Were mass and ether one or two? "Substance" also managed to divide itself into "like and unlike particles,"[40] "elementary particles endowed [another strange word] with the power of attraction and repulsion."[41] So Monism ran into the ancient *yin-yang* problem—two forces under one *Dao*, as Lao Zi said of old: "The Dao gave birth to one, and one to two, and two to three, and three to the ten thousand things," but still "the ten thousand things, the ten thousand forms, go back to one."[42] The age old conundrum of the one and the many, equally puzzling whether "the One" is "God" or "Substance."

But no matter. Haeckel wanted to be a Monist, and a materialist Monist, even though he was wary of the word materialism, noting quite rightly that "by the word '*materialism*,' two completely different things are very frequently confounded and mixed up."[43] He insisted, perfectly reasonably, that "as Darwinism, and in fact the whole theory of development, has been designated as 'materialistic,' I cannot avoid here at once guarding myself against this ambiguous word, and against the malice with which, in certain quarters, it is employed to stigmatize our doctrine."[44]

What he wanted to make perfectly clear was that Monism was not "moral materialism," materialism that "proposed no other aim to man in the course of his life than the most refined possible gratification of his senses."[45] And surely it was not. For in book after book, Haeckel held up as a cardinal tenet of his unconfessed faith "the profound truth that the real value of life does not lie in material enjoyment, but in moral action—that true happiness does not depend upon external possessions, but only in a virtuous course of life."[46]

Ironically, in guarding himself against the stigma of a materialism he despised, he destroyed his right to the materialism he worshiped. But we shall get to that. First, there was another form of materialism he decried, "the theoretical materialism that denies the existence of spirit, and dissolves the world into a heap of dead atoms."[47] For Haeckel, atoms were not "dead masses," but "living elementary particles."[48] That is why he was willing to call Monism "pantheism."

But in calling Monism pantheism, he did not give up his materialism. He could accept the pantheist's creed that "God and the world are one," because "in pantheism," he said, "God, as an *intra-mundane* being, is everywhere identical with nature itself, and is operative *within* the world as 'force' or 'energy.'"[49] He could accept pantheism because in the next breath he accepted Schopenhauer's gloss:

> Pantheism is only a polite form of atheism. The truth of pantheism lies in its destruction of the dual antithesis of God and the world, in its recognition that the world exists in virtue of its own inherent forces. The maxim of the pantheist, "God and the world are one," is merely a polite way of giving the Lord God his congé.[50]

And that, of course, is what Haeckel did: "The anthropomorphic notion of a deliberate architect and ruler of the world has gone forever. . . .The 'external, iron laws of nature' have taken his place."[51] And so the only fit philosophy was "scientific materialism":

> Scientific materialism, which is identical with our Monism, affirms in reality no more than that everything in the world goes on naturally—that every effect has its cause, and every cause its effect. It therefore assigns to causal law . . . its place over the entire series of phenomena that can be known. . . . Nowhere in the whole domain of human knowledge does it recognize real metaphysics, but throughout only physics.[52]

Haeckel did indeed want to be a materialist, and he did not shy away from the logical conclusions of materialism most at odds with our traditional sense of self. As a species we were an accidental product of a "mechanical process in which we discover no aim or purpose whatever."[53] As individuals we were each "but a tiny gram of protoplasm in the perishable framework of organic nature."[54] We were natural and nothing but natural. Haeckel here agreed with Huxley, "that man, physical, intellectual, and moral is as much a part of nature, as purely a product of the cosmic process, as the humblest weed."[55] Therefore even "all the phenomena of the psychic life" were "without exception, bound up with certain material changes in the living substance of the body, the *protoplasm*."[56] The vaunted "human mind," he said, "like that of any other animal, is a function of the central nervous system,"[57] and nothing more. Haeckel had no fear of what Chinese Marxists now disdain as "mechanical materialism." His "mechanical or monistic philosophy," he said, "asserts that everywhere the phenomena of human life, as well as those of external nature, are under the control of fixed and unalterable laws"[58] and that hence "every phenomenon has a mechanical cause."[59]

And so he came to what should have been his ultimate conclusion:

> The monism of the cosmos—proclaims the absolute dominion of "the great eternal iron laws" throughout the universe. It thus shatters, at the same time, the three central dogmas of the dualistic philosophy— the personality of God, the immortality of the soul, and the freedom of the will.[60]
>
> The freedom of the will. . . . is a pure dogma, based on an illusion, and has no real existence.[61]
>
> The great struggle between the determinist and the indeterminist, between the opponent and the sustainer of the freedom of the will, has ended today, after more than two thousand years, completely in favor of the determinist.[62]

With Goethe's help, Haeckel found our place in nature, dismissed the Riddle of the Universe, and put us in our place:

> By eternal laws
> of iron ruled,
> Must all fulfill
> The cycle of
> Their destiny[63]

Now there is nothing odd in that conclusion. Aside from those "living elementary particles," Haeckel's materialism was a quite conventional materialism quite conventionally reached. There was no neat trick in getting from "the course of the history of evolution" to mechanical materialism, the only possible materialism, it still seems to me, for a bona fide materialist. The neat trick was in getting from mechanical materialism to optimism.

Why should the protoplasmic puppets of the world rejoice? If free will—even severely limited free will—is an illusion, how is our existence not? If there is no action, only reaction, then where in the world are *we*? In the here and now we are not, if we act not. In the hereafter we will not be. With no will there is no way—even to pluck the day.

Haeckel did, it is true, try to cheer us up about the hereafter, though not, we must assume, of his own free will. He said that "the conception of a personal immortality cannot be maintained,"[64] but "immortality in a scientific sense" existed—in the great laws of the "conservation of substance," the "conservation of energy," the "conservation of matter."[65] In that, "we could take heart, for "at our death there disappears only the individual form in which the nerve-substance was fashioned."[66] Our energy marches on. Thus:

> Imperial Caesar, dead and turned to clay
> Might stop a hole to keep the wind away.
> O that the earth, which kept the world in awe,
> Should patch a wall t'expel the winter's flaw.[67]

Sic semper tyrannis. But cool comfort.

Haeckel tried again. We would not, he said, enjoy eternal life if we had it. For it would be *"endless."* Moreover it would not be exclusive: "It would be greatly marred by the prospect of meeting the less agreeable acquaintances and the enemies who have troubled our existence here below." Indeed, he added, (the cad), in an aside for which one can only hope his wife, if he had one, kicked him: "There are plenty of men who would gladly sacrifice all the glories of Paradise if it meant the eternal companionship of their 'better half' and their mother in law."[68] But was that cause for optimism?

Haeckel's logic failed him, but he failed to realize it. He did not get from monism to optimism step-by-step, but with a hop, skip, and a jump. By three leaps of faith he betook himself to the "palace of reason" where he proposed to "do reverence to the real trinity of the nineteenth century—the trinity of 'the true, the good, and the beautiful:'"[69]

> In the sincere cult of "the true, the good, and the beautiful," which is the heart of our new monistic religion, we find ample compensation for the anthropistic ideals of "God, freedom, and immortality" which we have lost.[70]

This was the clearest declaration he ever made of the source of his optimism—but it was riddled with riddles.

Beauty. In "this glorious wonderland of a world,"[71] Haeckel marveled at "the wonderful truths of the evolution of the cosmos" and at "the inexhaustible treasures of beauty lying everywhere hidden therein:"

> Whether we marvel at the majesty of the lofty mountains or the magic world of the sea, whether with the telescope we explore the infinitely great wonders of the starry heaven, or with the microscope the yet more surprising wonders of a life infinitely small, everywhere does Divine Nature open up to us an inexhaustible fountain of aesthetic enjoyment.[72]

For everywhere, as Darwin said himself, "endless forms most beautiful and most wonderful have been, and are being, evolved."[73]

Who would deny it? But how could we see it? "All the wonderful phenomena of nature," said Haeckel, ourselves included, "are only various products . . . of one and the same primitive matter."[74] But how could one product see beauty in other products? How could one "tiny gram of protoplasm," slave like all others to the "iron laws of nature," look at nature and see that it was good?

And was it good? Was it beautiful? Beauty was to be seen "in the green woods, in the blue sea, and on the snowy summits of the hills."[75] But was it also to be seen in "Nature red in tooth and claw,"[76] in the "relentless war of all against all," in "the raging war of interests in human society," that "is only a feeble picture of the unceasing and terrible war of existence which reigns throughout the whole of the living world?"[77]

Surely Haeckel was not blind to the struggle for existence, but he was able to look through it with the detachment of a *Chan* master:

> The modern man who "has science and art"—and, therefore, "religion"—needs no special church, no narrow, enclosed portion of space. For through the length and breadth of free nature, wherever he turns his gaze, to the whole universe or to any single part of it, he finds, indeed, the grim "struggle for life," but by its side are ever "the good, the true, and the beautiful." His church is commensurate with the whole of glorious nature.[78]

How did he manage it? One is reminded of Huxley's stark sentence: "The cosmic process is evolution; . . . it is full of wonder, full of beauty, and, at the same time, full of pain:"[79] "If anything is real, pain and sorrow and wrong are realities."[80] And the root of it all, for us, lay within us, in "the ape and tiger"[81] within us, in our "original sin," that would never go away, because "every child born into the world will still bring with him the instinct of unlimited self-assertion."[82] That is why Huxley, although he believed we could make things better, never succumbed to nineteenth century optimism.

But beauty blinded Haeckel to the beast—not just to Huxley's ape and tiger, but to the real grimness of life, to the real "pain and sorrow and wrong" that stand so starkly against the beauty of both art and nature. "Blind and insensible," Haeckel said, "have the great majority of mankind hitherto wandered through this glorious wonderland of a world; a sickly and unnatural theology has made it repulsive as a 'vale of tears.'"[83] To each his own blindness. The absolutely mysterious beauty of nature gives comfort beyond comprehension. But by whose abacus can it simply add up to "compensation?" And can it alone dry all tears?

Goodness. Haeckel's pronouncements on "goodness" were riddled with contradictions. It was already, perhaps, mildly contradictory for one so bent on awakening us to the "unavoidable conflict between the discredited dominant doctrines of Christianity and the illuminating, rational revelation of modern science"[84] to urge us to turn to Christianity for "our Monist ethics," but that is what he did. "We must endeavor," he said, "to save from the inevitable wreck of this great world religion for our new monistic religion . . . the principles of true humanism, the golden rule, the spirit of tolerance, the love of man." It helped, of course, to note in the next breath that "these true graces of Christianity were not, indeed, first discovered and given to the world by that religion, but were successfully developed in the critical period when classical antiquity was hastening to its doom,"[85] but once that was clear, Haeckel had no qualms about claiming that "our monistic ethics is [sic] completely at one with Christianity," at least when it came to Christianity's "greatest and highest commandment," "thou shalt love thy neighbor as thyself."[86] But as soon as he interpreted that commandment, it was clear that monistic ethics were not at one with Christianity at all, and could not be, because Christianity had got its own highest commandment all wrong: "Christian ethics was [sic] marred by the great defect of a narrow insistence on altruism and a denunciation of egoism".[87]

> The supreme mistake of Christian ethics and one which runs directly counter to the Golden Rule, is its exaggeration of love of one's neighbor at the expense of self-love. Christianity attacks and despises egoism on principle. Yet that natural impulse is absolutely indispensable in view of self-preservation; indeed, one may say that even altruism, its apparent opposite, is only an enlightened egoism. Nothing great or elevated has ever taken place without egoism.[88]

So it was up to Monism, not Christianity, to show the way: "The highest aim of all ethics is very simple—it is the re-establishment of 'the natural equality of egoism and altruism, of the love of one's self and the love of one's neighbor.'"[89] Being totally reasonable, "our monistic ethics lays equal emphasis on the two and finds perfect virtue in the just balance of love of self and love of one's neighbor."[90]

Here again we see Haeckel's breathtaking ability to stare at contradictions and not see them. How in an either/or world could one strike "a just balance" between opposites? How, in Evolution's name, could there be a *natural* equality of egoism and altruism? Or if there was one, how could it require "re-establishment?"

Haeckel said it was all perfectly natural: "Man belongs to the social vertebrates, and has, therefore, like all social animals, two sets of duties—first to himself, and second to the society to which he belongs." Did that "first" destroy the balance? Somehow no: "The former are the behests of self-love or egoism, the latter of love for one's fellows or altruism. The two sets of precepts are equally just, equally natural, and equally indispensable."[91] So *que faire?* Egoism on Monday, Wednesday, and Friday, and altruism on Tuesday, Thursday, and Saturday? And then on one's day of rest, when forced to save oneself, or one's fellows, death for them and *seppuku* for oneself, all with Solomon's sword?

What Haeckel needed was the monist sociobiologists' "selfish gene,"[92] which could egoistically order the altruistic sacrifice of one carrier to save two. But he did not yet know of selfish genes. He still seemed to think that egoism meant self-preservation for the individual.

When, in the "struggle for life," race fought with race, racial egoism tipped the balance. We can see that in Haeckel's ridicule of another Christian virtue:

"If any man will take away thy coat, let him have thy cloak also." Translated into the terms of modern life, that means: "When some unscrupulous scoundrel has defrauded thee of half thy goods, let him have the other half also." Or again, in the language of modern politics: "When the pious English take from you simple Germans one after another of your new and valuable colonies in Africa, let them have all the rest of your colonies also—or, best of all, give them Germany itself."[93]

But what we were to do when our "duties" to self and race came into conflict, Haeckel failed to say.

He also failed to say what "duties" were doing in evolution in the first place, although he tried: "The social duties which are imposed by the

social structure of the associated individuals, and by means of which it se-
cures its preservation, are merely higher evolutionary stages of the social
instincts, which we find in all higher social animals (as 'habits which have
become hereditary')."[94] If we ignore the "hereditary habit" problem, an
echo of Darwin in a Lamarckian lapse,[95] it is still problematic enough that
Haeckel should find duty, virtue, and instinct all the same thing:

> The fidelity and devotion of the dog, the maternal love of the lion-
> ess, the conjugal love and connubial fidelity of doves and love-birds
> are proverbial, and might serve as examples to many men. If these
> virtues are to be called "instincts," then they deserve the same name
> in mankind.[96]

But instincts are *not* virtues. If all is instinct, there *is* no virtue. If all is
natural law there is no duty. If all is natural law there is no moral law. If
there is no free will, there is no disobedience.

And so we come to the ultimate contradiction in the first and great
commandment for all good Monists, that Haeckel gives us at the very end
of his *History of Creation*:

> It is above all things necessary to make a complete and honest re-
> turn to Nature and to natural relations. —As Fritz Ratzel has excel-
> lently remarked, "[We must] . . . endeavor to lead a life according to
> natural laws."[97]

But how can we do anything else! If "the great, eternal iron laws"
hold "absolute dominion . . . throughout the universe," by what fall is
man free to break them? "Monism . . . shatters . . . the freedom of the
will." "We have lost" our freedom. But if we have lost our freedom, we are
free from all moral responsibility.

Truth. We return to nineteenth-century know-it-allism: All the truth
worth knowing could be known—to our little mechanical minds:

> The highest function of the human mind is perfect knowledge, fully
> developed consciousness, and the moral activity arising from it.
> "Know thyself!" was the cry of the philosophers of antiquity to their
> fellow-men who were striving to ennoble themselves. "Know thy-
> self!" is the cry of the Theory of Development, not merely to the in-
> dividual but to all mankind. . . . Mankind as a whole will be led to a
> higher path of moral perfection by the knowledge of its true origin
> and its actual position in Nature.[98]

Let us ignore, for a moment, the metaphysical mystery of "moral perfection." How could *we* have "perfect knowledge?" How could any "tiny gram of protoplasm," any chance concurrence of "living atoms" have "perfect knowledge?" How could any *thing*, any creature of God or Nature, eat of the fruit of the tree of knowledge? That riddle defied "the philosophers of antiquity." It defies their descendents. No theist, atheist, materialist, or idealist has ever answered it. Indeed, had that riddle been the riddle of the Sphinx, the Sphinx would still be in business, still alive and well, and very well fed, replete ("I repeat, ree-plete"[99]) with the flesh of every possible flavor of philosopher. But Haeckel, perfect in his conceit, once again failed to recognize a riddle when he saw one.

And so he felt free—tiny, unfree gram of protoplasm that he said he was—to dispense the truth, to propagate "true knowledge" of Nature,[100] which meant, naturally, knowledge of the Law of Substance, knowledge of the Theory of Development, and knowledge of "the great law of progress or perfecting."[101]

In "the real trinity of the nineteenth century—the trinity of 'the true, the good, and the beautiful,'" he found a second trinity. In truth, he found substance, evolution, and progress, three more in one, one still in three. Or one in five? Or one in six? Monist mathematics? Buddhist mathematics? Taoist mathematics? Anyway, another trinity *that did not, does not, and will not add up—without a measure for "progress."*

But that is the rub. Whence came Haeckel's measure for progress, his measure of his "great law of progress or perfecting?" Was it in the Law of Substance? *No.* For the Law of Substance was the Great Equalizer, the Great Reducer of all things—to matter, and nothing else. A kaleidoscopic cosmos full of glass. Or dust: From dust to dust, and the stuff of dust in between (dead or alive, our stuff is stuff). Or mist: "For you are mist, appearing for a while, and then disappearing."[102] Metaphor earthy or ethereal, it is all the same. Form is ephemeral. Matter is matter. How can there be progress?

Then is "the great law of progress and perfecting" part of "the Theory of Development," the theory of evolution? *No.* For the theory of evolution has no way to measure progress either. It is an explanation of "descent with modification"—and of nothing else.[103] It relates all things, but it does not rank them. It distinguishes the fit from the unfit, but it does not distinguish *among* the fit, for fitness is its only value, and fitness is nothing but the genetic ability (the fruit of genetic luck) to be fruitful and multiply. How creatures multiply is of no account. "Nature," says Darwin, "cares nothing for appearances."[104] Fitness has ten thousand faces, all equal in the eyes, blind eyes, of Natural Selection, which holds the gate with the impartiality of a Legalist Sage-King. Nature puts each of the ten thousand things to the same test: Be fruitful and multiply. Nature gives laurels to all who pass it, whatever they look like, whatever they do in their spare time.

In form, Natural Selection separates the sheep from the goats. But in "worth," whatever that is, Natural Selection does *not* separate the sheep from the goats. *We* do. *We* sit on the sidelines, applauding this and deploring that. *We* look at evolution and declare it good. *We* look at evolution and see in it improvement, reaching its height in us. But the standards by which we so judge ourselves and our fellow creatures are *not* to be found in *The Origin of Species*. This is the truth we must grasp if we are to "evaluate" (how we evaluate anything is the mystery) the views of evolution of Haeckel and Lu Xun: In *The Origin of Species* Darwin gives us good cause to believe in evolution. He gives us *no* cause to believe in progress.

But—someone should protest—Darwin himself believed in progress. And he did speak of progress and perfection in *The Origin of Species* over and over again.

He did indeed. But in every case he was at his own throat.

Consider again his most seductive sentence on progress: "As natural selection works solely by and for the good of each being, all corporeal and mental endowments will tend to progress towards perfection."[105] How consoling! All those dactyls. All that alliteration. Progress inevitably marching on. If natural selection be for us, who can be against us? If Darwin says so . . .

But Darwin's sentence *makes no Darwinian sense*—for six reasons:

1. Natural selection does not work "for the good of each being." The unfit do not "get what's good for them," unless extinction is indeed Nirvana, unless the breaking of the transmigration of the gene is the ultimate act of compassion, unless the truth lies in biological Buddhism. Natural selection does not work for the good of the unfit, unless there is some ineffable ecstasy in genetic dissolution.

    But natural selection does not work for the good of the fit either. For natural selection does not do anything to the fit. It does not *make* the fit fit. It works no change on any being. Darwin wrote, in the introduction to *The Origin* that "any being, if it vary however slightly in any manner profitable to itself, under the complex and sometimes varying conditions of life, will have a better chance of surviving, and thus be *naturally* selected."[106] But *no being* does vary genetically, even slightly, in its lifetime. No being adapts itself to its environment. Each being is genetically set at its conception. We shall be changed, but not genetically and not by natural selection. "Variation" occurs *between* generations. It comes occasionally from mutation in the production of our sex cells. It comes inevitably in the chromosomal mixing mandated by sexual reproduction. But none of that is natural selection's work. Natural selection does nothing but make descent with modification obvious over time.

2. Natural selection does not work for the good of each species. Natural selection is not humane. It treats the ten thousand species like straw dogs.[107] It "allows" countless species to die out, without issue. And even when there is issue, when species "are survived" by modified descendants, they are not saved by those descendants. Long-necked giraffes did not save their short-necked progenitors. They saved some of their genes, many of their genes, but not those genes that "distinguished" them, otherwise short-necked giraffes would not have perished from the face of the earth. Life has gone on from the beginning, despite "the war of nature."[108] New forms have come out of old—new species have evolved—but they have not preserved the old.

3. Natural selection is *not* engaged in "the work of improvement."[109] It may *not* be said "that natural selection is daily and hourly scrutinizing, throughout the world, every variation, even the slightest; rejecting that which is bad, preserving and adding up all that is good, silently and insensibly working, whenever and wherever opportunity offers, at the improvement of each organic being in relation to its organic and inorganic conditions of life."[110] All that may be said is that natural selection, when the organic and inorganic conditions of life throw organic beings into competition, silently and insensibly "selects" those beings most fitted to those conditions. But the fittest are usually *not* the modified. Mutants, after all, are freaks, and freaks, generally speaking, have a hard time of it. It is only when there is some freakish change in the environment that they have their day.

Darwin knew that, of course. Arguing that "favourable variations" would be preserved, he knew that unfavorable variations would not.[111] But he was obsessed with explaining change, and so naturally said little about the conservative force of reproduction and natural selection. His task, after all, was to explain how change could occur despite that force. And he was perfectly right. Change depended on "profitable variation": "Unless profitable variations do occur, natural selection can do nothing."[112] But what could it do when they did occur? It could only do what it always did: choose fitness—new fitness, but still just fitness.

Darwin kept talking of species that were "modified and improved."[113] But what did he mean by "improvement?" Was there "improvement" in the "divergence of character"[114] he found among the finches of the Galapagos? No. He found no qualitative great leap forward in finchitude. He found no *Überfink*. Viable divergent forms did not outfinch each other. Not even the thorn-wielding cactus finch can be said to have swashbuckled its way to the top of the finch family tree, because there was no top, because there was no set direction to finch evolution. Bigger beaks were not necessarily better beaks. Bigger beaks proved better on some islands. Finer beaks proved fitter on others. Indeed, beaks could wax and wane in a single population on a single island. If evolution went anywhere, it went in different directions at once. So which was progress?

Darwin wrote of the Galapagos finches in *The Voyage of the Beagle,* in the sentence that came closest to his eventual theory of the origin of species, that "one might really fancy that from an original paucity of birds in this archipelago, one species had been taken and modified for different ends."[115] But that is not really what happened. There were no ends. Different islands favored different forms. Under different conditions certain mutants chanced to prove fitter than others—and fitter than nonmutants, the genetically filially pious. On each island, natural selection produced "perfection," said Darwin, "or strength in the battle for life, only according to the standard of that country."[116] Darwin's term "perfection" was relative and limited—but even so misleading.

The limitations and relativity of "perfection" and "improvement" are clearly seen in the famous Birmingham pepper moth case. Once upon a time, white moths ruled the woods, because they were well camouflaged on clean trees. Black mutants never multiplied, because they were too easily picked off, by birds not yet endowed with eagle eyes. But then the industrial revolution reversed things. In befouled forests, white moths had no place to hide, so they fed the birds while the mutants mated—and multiplied. So, pepper moths "turned black"—until in recent times antipollution efforts let white moths make a comeback. From white to black from black to white, was white or black improvement?

The moral is that improvement is not a matter of black and white. Fitness too is ephemeral. The fit of today may be the forlorn of tomorrow and the forgotten of the next day. Specification may be a prelude to disaster. Who can say what is improvement? Who can say what is progress? Who can say what is perfection?

4. Darwin meant by "perfection of parts," that is, the perfection of the function of parts, the fitting of organs to environments. But even here the word "perfection" dissolves in relativity. Let us look at the eye (an amazing trick), that organ above all others that stood, in Darwin's own eyes, as a challenge to the creative powers of natural selection. "[The] eye," Darwin wrote in 1860 to Asa Gray, "to this day gives me a cold

shudder, but when I think of the fine known gradations, my reason tells me I ought to conquer the cold shudder."[117] He could conquer it because he could understand the eye evolving over eons. But did "the eye" evolve? Eyes evolved, all sorts of them: fish eyes, eagle eyes, cats' eyes, our eyes. But *the* eye? The perfect eye? Where is that? To each its own? Then some eyes are "perfectly" all right although they hardly see. The Yangtze River dolphin's eyes are almost out of sight, and that is *not* the cause of the Yangtze River dolphin's near extinction. It is man not mud that has fouled its niche. The Yangtze River has said, "Here's mud in your eye" for millennia, and the Yangtze River dolphin has said, "Who cares?" For mud selected sonar—or sonar conquered mud. So the Yangtze River dolphin's eyes receded, *not* through "disuse," but because little, useless eyes proved useful—less open to disease. So natural selection can "perfect" in opposite directions.

5.   But really natural selection goes in no direction. It cannot "work towards" anything, because it has no mind's eye. It does not know where it is going. It is not going anywhere. It is not up to anything. There are no Platonic forms, no perfect eyes, no perfect ears, no perfect anything toward which it aspires. It is only in looking backward, into fossil records, that we see "progressions," but they are *not* progressions. We see the paths by which strange things have come. We see no paths that any creature took, because everything that has ever lived has lived an aimless existence. Evolution is aimless—unless there is more to evolution than meets the eye, and that is what Darwin sought to disprove. But he could not disprove it with talk of progress and perfection.

6.   When Darwin said, "All corporeal and mental endowments will tend to progress toward perfection," he was not, to be fair, talking about progress toward a perfect being. In that passage, he was not talking about a great chain of being at all, but about parts. But elsewhere he gave himself away. He too believed that evolution's family tree marked progress. He said in the last sentences of *The Origin* that "the most exalted object of which we are capable of conceiving" was "the production of the higher animals." And then, imagining the "few forms" or "one," from which that production had begun, he marveled that "from so simple a beginning endless forms most beautiful and most wonderful have been and are being evolved."[118]

    What a marvelous sentence. Progress from the simple to the wonderful. How much more exciting than Spencer's prosaic progress from the simple to the complex—or from "the homogeneous to the heterogeneous!"[119] But Darwin's *theory* proved no such progress. If it demonstrated anything, it demonstrated the opposite, that there were no meaningful distinctions between simple and complex, that there were, indeed, no "higher" or "lower" animals, that all forms were equal as long as they were fit, that Evolution's motto, if it had one, could only be, "If life goes on, who cares how?" Darwin admitted it. "Maternal love," he said, "or maternal hatred . . . is all the same to the inexorable principle of natural selection." Granted, he added parenthetically, "though the latter fortunately is most rare," but his theory had no way of explaining that "fortunately."[120]

For—let us say it once again—*Darwin's theory has only one value:* fitness, and fitness can come in any guise: in simplicity, in complexity, in strength, in weakness, in brain, in brawn, in love, in hate. There is only one law, "one general law . . . namely, multiply, vary, let the strongest live and the weakest die"[121]—and anything that survives is "strong." One law, and all are equal before the law, all fit until proven unfit.

So again, where is there room for "progress?" Only, in theory, in fitness itself. But have species become fitter and fitter? Surely not for sure. The fittest things we know are simple things compared to us. Our "intelligence," whatever that is, makes us, we think, the most complex thing the earth has ever known, but we have not survived here long enough to be the *fittest* thing the earth has ever known. We sometimes say we are, because of our ability to fit in anywhere on earth, even, in artificial niches, beneath the sea, or in the sky, or—the ultimate trick up our species' sleeve—across the vast reaches of space, on other worlds. But the power that makes that possible is the very power that may make it impossible. We do not know. We may blast off to colonize other worlds (even then we might take things with us that could outlive us) or we may blast ourselves off the face of the earth—to smithereens and nowhere else. We cannot know. Who is the fittest of them all? There is no mirror on the wall:

> The brontosaurus had a brain
> No bigger than a crisp.
> The dodo had a stammer,
> And the mammoth had a lithp.
> The auk was just too awkward,
> Now they're none of them alive.
> Each one, like man, had shown himself
> Unfitted to survive.
> Their story points a moral,
> Now it's we who wear the pants.
> The extinction of these species
> Holds a lesson for us ants.[122]

Have lessons effected natural selection? Who will laugh last? Darwin, alas, did not prove that we will.

So Haeckel's "great law of progress and perfecting" was not Darwinian. It was not Monist. It was not materialist. It was idealist to the core. His insistence on the existence of such a law was an expression of an unrecognized idealism as irrepressible as Dr. Strangelove's salute.

"The development of the universe," Haeckel the Monist said, "is a monistic mechanical process, in which we discover no aim or purpose whatever."[123] "On the whole," Haeckel the idealist said, "the movement of development of all mankind is and remains a progressive one, inasmuch as man continually removes himself further from his ape-like ancestors, and continually approaches nearer to his own ideal."[124]

But how in evolution could we do that? Were we even more amazing than Emerson's worm?

> And striving to be man the worm
> mounts through all the spires of form.[125]

Evolutionary low life worming its way into high society? That was silly. But Haeckel's version of evolution was even more so. He had us in the worm. He had us striving to mount through all the spires of form, striving to be, before we were, striving to slough off wormitude and every other beastly guise that came before us in our family line, striving to reveal ourselves at last in our *true* form, our *ideal* form. That was an idealism that left Plato in the dust. That was biological Buddhism.

Through countless previous incarnations, we worked out our evolution with diligence, winning ourselves our "manhood" (*again pace feminae*). Haeckel, in yet another stunning contradiction, gave us credit for our evolution:

> Our Theory of Development explains the origin of man and the course of his historical development in the only natural manner. We see in his gradually ascensive development out of the lower vertebrata, the greatest triumph of humanity over the whole of the rest of Nature. We are proud of having so immensely outstripped our lower animal ancestors.[126]

But we deserve no credit for our evolution. We had nothing to do with it. Individuals struggle to survive—and mate; species do not struggle to evolve. *We* did not gradually ascend in triumph out of the lower vertebrata. We were not there to outstrip our lower animal ancestors. Creatures "triumph" only over others in their generation—and even then their triumphs are not triumphs, for in evolution, victory is in the genes—or blind luck (bolts from the blue fall equally on the fit and the unfit). In evolution all fights are fixed—because the genes are fixed. "The environment" may throw a fight to this set of genes or to that, but every set is set. Mutations are *faits accomplis* at conception. So humanity owes all to inheritance and good fortune. Of what can we be proud?

Any way we look at it, if we are to explain our origin in a natural manner, Darwin gives us no cause for shame—and no cause for pride.

But Haeckel could not conceal his pride. He could not conceal an unDarwinian anthropocentric pride, even though he took pride in decrying anthropocentric pride in others. He railed against "anthropism," against "the anthropocentric dogma" that "culminates in the idea that man is the preordained centre and aim of all terrestrial life," against "the prevalent illusion of man's supreme importance, and the arrogance with which he sets himself apart from the illimitable universe, and exalts himself to the position of its most valuable element."[127] And then he proclaimed the truth of the "amply demonstrated" "cosmological theorem" that "the most perfect and most highly developed branch of the class mammalia is the order of primates" and "the youngest and most perfect twig of the branch primates is man."[128]—"What is man that thou art mindful of him?" "Man is the measure of all things."

Had it really been "amply demonstrated that man was *wan wu zhi ling* (the best of the ten thousand things)? Of course not. But Haeckel honestly believed that it had. He did not simply claim that "truth" to be self-evident. Despite his denunciation of the biblical doctrine of the primacy of man, he insisted that there was scientific proof of a law of progress that led to man—in two "intimately connected" branches of natural history: phylogeny and ontogeny, the former referring to the evolutionary history of a species and the latter to the biological development of an individual.[129]

Proof number one had been discovered by paleontologists: The "great and important" "law of progress" had "long been empirically established by paleontological experience, before Darwin's Theory of Selection gave us the key to the explanation of its cause. The most distinguished paleontologists have pointed out the law of progress as the most general result of their investigations of fossil organisms."[130]

Proof number two, even "more wonderful," even "more mysterious,"[131] Haeckel had discovered himself, as the result of his own investigations of—wonder of wonders—*unborn* organisms, as the result of his investigations in embryology.

Granted, others may have glimpsed the truth, others may have helped to establish empirically the phenomena that formed the basis of this second proof, but only he, with Darwin's key, had truly unlocked the secret of embryology, of ontogeny. Only he had truly discovered "the first principle of Biogeny," yea, "the fundamental law of organic evolution." Only he had fully understood "that Ontogeny is a recapitulation of Phylogeny."[132]

That was Haeckel's famous, fallacious law, a "law," as Stephen Jay Gould has pointed out, of most unfortunately "pervasive influence," in the fields of criminal anthropology, child development, primary educa- tion, and Freudian psychoanalysis, as well as in the as yet undispellable miasma of racial, and sexual, prejudice,[133] an influence pervasive enough

to reach Lu Xun—to whom, fear not, we will return.

Haeckel's law was not ridiculous. He leapt to a conclusion to which any of us might have leapt. For embryology in its infancy was an amazing science. It still is an amazing science. The journey from human zygote to human infant is an amazing journey, and it is not surprising that Haeckel should find it amazing that along that journey we should exhibit, for a while, embryonic gills and embryonic tails. The trouble was that Haeckel did not say that we exhibited embryonic gills and tails. He said we had gills and tails themselves. He assumed that early on, the human embryo had indeed "essentially the anatomical structure of a Lancelet, later of a Fish, and in subsequent stages those of Amphibian and Mammal forms; and that in the further evolution of these mammal forms those first ap- pear which stand lowest in the series, namely, forms allied to the Beaked Animals *(Ornithorhynchus);* and then those allied to Pouched Animals *(Marsupialia),* which are followed by forms most resembling Apes; till at last the peculiar [*sic*] human form is produced as the final result."[134]

In short he assumed that "the embryonic development of the indi- vidual is completely parallel to the paleontological development of the whole tribe to which it belongs."[135] That is what it meant to say that "On- togeny is a recapitulation of Phylogeny":

> The series of forms through which the Individual Organism passes during its progress from the egg cell to its fully developed state, is a brief, compressed reproduction of the long series of forms through which the animal ancestors of that organism (or the ancestral forms of its species) have passed from the earliest periods of so-called or- ganic creation down to the present time.[136]

There were complications. Although Haeckel insisted that the two series were "completely parallel," he admitted that the complete phyloge- netic series was rarely repeated in the ontogeny of an ordinary individual of any species. The ontogenetic series was "compressed": there were missing links. Nonetheless the "sequence" was the same.[137] One might miss this branch or that branch, but every being had to climb its family tree to be. Every mammal had to climb its family tree to get out of the womb. Each was *forced* to, because "Phylogeny is the mechanical cause of Ontogeny."[138]

Was it? No. We do not literally evolve in nine months the way our species evolved over all the wastes of time since the beginning. We do start from a single cell, as do all mammals, but those cells are different, each already coded in DNA to grow into a unique individual, freakish or unfreakish, of an existent species.[139] Nothing is ever born totally new under the sun. The admittedly remarkable similarity of vertebrate embryos does not prove that we all start from scratch and climb flipper over flipper, hoof over hoof, hand over hand up the evolutionary tree to our appointed branch. Haeckel did have good reason to see in embryology confirmation of the Theory of Descent.[140] Darwin explains embryology better than anyone else. It is hard to imagine that God would have created such a show just for the amusement, millions of years after the show began, of people who cut up fetuses. But the Theory of Descent itself provides a better explanation (partial not complete) of embryology than Haeckel did. Our embryonic gills and tails merely reflect what Gould has called, in a line of argument that goes all the way back to von Baer, "a conservative principle of heredity."[141] The single-celled method of sexual reproduction has proved eminently successful. Gill slits, which got into that process early in the day, can still be seen even in our embryos, for a while, either because, for a while, they still perform some useful function, *or* because they have never gotten in the way of our evolving DNA—and so have never been selected out.

"The connection of ontogeny and phylogeny" was not as "intimate" as Haeckel thought.[142] Nor was it as neat. And yet it was the neatness of that seeming intimacy that gave his "law" its seductive and destructive power—precisely, alas, because it seemed to prove progress. Surely, far too many of his readers thought, if he had found the same series in fossils and fetuses, he must *have* something. But what did he have? Even had he been right about the "intimate connection of ontogeny and phylogeny," he would not have had what he thought he had.

He thought he had discovered in both phylogeny and ontogeny one and the same "series of progressive steps," steps that marked "the different degrees of perfection of the divergent branches of the tribe."[143] But he had not. All he had established was a phylogenetic line of descent or an ontogenetic line of development, neither of which said a thing about human "perfection" or human "worth." *Of course* if one started with man and looked for man's antecedents, one found roads that led to man, but all roads did not lead to man. Phylogenetic and ontogenetic lines led to man, but phylogeny and ontogeny did not. Phylogeny and ontogeny led to every living species, and by lines of *identical length*. We were not "the youngest and most perfect twig" at *the* top of the tree of evolution. For all living species were at the top. The top of evolution's tree was flat. No, the top was not flat: there was no top. "The great Chain of Being" still snaked up Haeckel's evolutionary tree,[144] but in truth there was no great chain of being. For in truth there is no tree. Evolution is a ground cover. It is a vine, of one root, of a million offshoots, with offshoots of offshoots—millions of which have withered and died, millions of which live on. We are at the end of an offshoot. Every extant species is at the end of an offshoot, each and every one equidistant from the Root.

Haeckel was fooled by his tree of life, fooled by his faith that the last shall be first. For among living species there is no last. All living species are of the same generation. We are cousins of all extant species, sons and daughters of none.

Only on our own offshoot of evolution's vine are we the last, or the latest. But even there, there is no cause to be sure that the last shall be first. Our offshoot may live. Our offshoot may die. In either case our segment of it might someday prove longer than any that has gone before, but it is not so yet and it may never be. For no species can know the measure of its days. What we can know is that not even by virtue of our intelligence are we necessarily the fittest of them all. We may prove wise enough to live happily ever after. We may prove too smart for our own good.

At any rate there is *no* evolutionary way to declare us unequivocally the crowning ornament on an evolutionary tree. If, with Haeckel, Huxley, Bishop Wilberforce, and Darwin himself, we insist that we are, we *ought* to admit that we do so for extra-evolutionary reasons.

But Haeckel did not. He protested to the end that evidence of evolution was proof of progress. The true source of his optimism was anthropocentric (and German) self-satisfaction, but he thought it was his "true knowledge" of the "Doctrine of Descent":[145]

We may justly hope . . . that the progress of mankind towards freedom, and thus to the utmost perfection, will by the happy influence of natural selection, become more and more certain.[146]

What heroic illogic, for a monist, materialist, determinist. But materialist or idealist, he was wrong: Evidence of evolution is *not* proof of progress. Progress is a metaphysical concept.

And yet Haeckel's hope was such a "nice" one, so encouraging, and so common. Why should it hurt people?

Haeckel's "law of progress" hurt people in two ways. The first was philosophical. He gave people false cause for optimism. There may be good cause, but his was not it.

The second was more down to earth: Haeckel's "law of progress," bolstered by his "biogenic law," confirmed for many—many too many, a "natural" belief in a natural hierarchy of species—and races.

Having found in both phylogeny and ontogeny a "series of progressive steps" that marked "the different degrees of perfection of the divergent branches of the tribe," he easily declared that human races represented different degrees of perfection of the divergent branches of the human tribe. But that meant that there really was no tribe. Human races, he said, were different species:

> All . . . races of men, according to the Jewish legend of creation, are said to be descended from "a single pair,"—Adam and Eve, —and in accordance with this are said to be varieties of one kind or species. If, however, we compare them without prejudice, there can be no doubt that the differences of these five races are as great and even greater than the "specific differences" by which zoologists and botanists distinguish recognized "good" animal and vegitable species ("bonae species"). The excellent paleontologist Quenstedt is right in maintaining that, "if Negroes and Caucasians were snails, zoologists would universally agree that they represented two very excellent species, which could never have originated from one pair by gradual divergence."[147]

Haeckel had not abandoned Darwin's theory of divergence. He assumed "a *single primaeval* home for mankind, where he developed out of a long since extinct anthropoid species of ape."[148] He assumed that from that species of ape had evolved a single species "of speechless primaeval man, whom we consider as the common primary species of all the others." But those "others" were "various species . . . who still remained at the stage of speechless ape-men."[149] And it was from those "various species," not from a common *human* ancestral species, that our modern "races" had evolved—unto "different degrees of perfection."

Haeckel threw all the weight of his "science" into distinguishing those degrees. A "systematic survey" of "genuine or talking men" revealed—to Haeckel—"twelve species of men" and "thirty-six races," in three great groups, distinguished first by language and then—by hair, the texture of which made their "different degrees of perfection"—to Haeckel—perfectly obvious.[150]

"Of the twelve species of men," he said, ". . . the four lower species are characterized by the wooly nature of the hair of their heads." These were the Papuans, Hottentots, Coffres, and Negroes.[151] Then there were "the eight higher races of man," the "straight haired" species, in which "the hair of the head is never actually wooly, although it is very much frizzled in some individuals." But these higher straight-haired species could be divided into two groups: "stiff-haired men, —Australians, Malays, Mongolians, Arctic Tribes, and Americans," and "curly-haired men—the Dravidas, Nubians, and Mediterranean races."[152]

It was not difficult to discern Haeckel's favorites. "All Ulotrichi, or wooly-haired men," said he, "have slanting teeth and long heads, and the color of their skin, hair, and eyes is always very dark. ["The color of the skin of genuine negroes is always more or less of a pure black. Their skin is velvety to the touch, and characterized by a peculiar offensive exhalation."][153] They are on the whole at a much lower stage of development, and more like apes, than most of the Lissotrichi, or straight-haired men."[154] Indeed, "considering the extraordinary resemblance between the lowest wooly-haired men, and the highest man-like apes, which still exist at the present day," he said, "it requires but a slight stretch of the imagination to conceive an intermediate form connecting the two, and to see in it an approximate likeness to the supposed primaeval men, or ape-like men."[155] (Elsewhere, to help those without such imagination, he provided a plan showing equally imaginative portraits of a chimpanzee, gorilla, "orang," and naked "negro," each perched in the branches of a tree!).[156] If one still had doubts, one had only to note the telling evolutionary "fact" that "both the African Man-like Apes are black in color, and like their countrymen, the Negroes, have the head long from back to front (dolichocephalic). The Asiatic Man-like Apes are, on the contrary, mostly of a brown, or yellowish brown color and have the head short from back to front (brachycephalic), like their countrymen, the Malays and Mongols."[157] Such was the science of the day.

But the worst part of his argument (in every sense of the word) still a part of his tree argument, was that the "wooly-haired men" were forever out on an evolutionary limb:

> The Ulotrichi are incapable of a true inner culture and of a higher mental development, even under the favorable conditions of adaptation now offered to them in the United States of North America. No wooly-haired nation has ever had an important "history."[158]

Nor would one, at least none of the wildest:

> All attempts to introduce civilization among [the wildest tribes in southern Asia and eastern Africa], and many of the other tribes of the lowest human species, have hitherto been of no avail; it is impossible

to implant human culture where the requisite soil, namely, the perfecting of the brain, is wanting. Not one of these tribes has ever been ennobled by civilization; it rather accelerates their extinction.[159]

Haeckel's science had descended to the evidence of frustrated missionaries:

> Even many Christian missionaries, who, after long years of fruitless endeavors to civilize these lowest races, have abandoned the attempt, express the same harsh judgment, and maintain that it would be easier to train the most intelligent domestic animals to a moral and civilized life, than these unreasoning brute-like men.[160]

And it was on the very next page that he said, "The Theory of Descent as applied to man opens up the most encouraging prospects, for the future."[161]

The future was not even bright, however, for all of the "straight-haired" races:

> The lowest stage of all straight-haired men, and on the whole perhaps of all the still living human species, is occupied by the *Australian*, or *Austral-negro* (Homo Australis). This species seems to be exclusively confined to the large island of Australia; it resembles the genuine African Negro by its black or brownish black hair, and the offensive smell of the skin, by its very slanting teeth and long-headed form of skull, the receding forehead, broad nose, protruding lips, and also by the entire absence of calves.[162]

It is a pity Haeckel had not heard of what Chinese, and I assume Africans and Australians, thought, and think, of the fragrance of sweaty Europeans. Soap has cleaned up a lot of racial prejudice.[163]

Haeckel had little to say about the Chinese. (Perhaps, for Lu Xun's sake, it was just as well.) They were a middling race of a middling species, "the Mongols." He granted the species a history, but not the Darwinian fitness prize (Darwin's only prize) that it deserved ("the Mongol [Homo Mongolus] is, *next to the Mediterranese*, the richest in individuals"). He commented on color ("The color of the Mongol is always distinguished by a yellow tone, sometimes a light pea green [*sic*], or even white, sometimes a darker brownish yellow"), hair, skull, and general appearance ("The narrow openings of their eyes, which are generally slanting, their prominent cheek bones, broad noses, and thick lips are very striking"), and then, struck by little else, he let the "Mongols," and Chinese, alone.[164]

His interest lay elsewhere, among "the curly-haired," and there, among "the Mediterranese" (The Dravidas and Nubians, members of the group thanks chiefly to their sharing in "the strong development of the beard," were given short shrift):[165]

> The *Caucasian,* or *Mediterranean man* (Homo Mediterraneus), has from time immemorial been placed at the head of all races of men, as the most highly developed and perfect. . . . In bodily as well as in mental qualities, no other human species can equal the Mediterranean. This species alone (with the exception of the Mongolian) has had an actual history; it alone has attained to that degree of civilization which seems to raise man above the rest of nature.[166]

Haeckel surely believed that he was belaboring the obvious:

> The characteristics which distinguish the Mediterranean from the other species of the race are well known . . . it is only in this one species of men that the body as a whole attains that symmetry in all parts, and that equal development, which we call the type of perfect human beauty.[167]

But even within this exalted group, evolution had made distinctions, and here Haeckel's denial of common human ancestry for all races proved most useful: "Semites and Indo-Germani are descended from different ape-like men."[168] Needless to say it was the "Indo-Germanic race, which has far surpassed all the other races of men in mental development,"[169] that had "deviated furthest from the common primary form of ape-like men,"[170] but even it had branches. As Haeckel's argument wound down, it was clear where he was going:

> During classic antiquity and the middle ages, the Romanic branch (the Graeco-Italo-Keltic group), one of the two main branches of the Indo-Germanic species, outstripped all other branches in the career of civilization, but at present the same position is occupied by the Germanic. Its chief representatives are the English [a bow to Darwin?] and Germans, who are in the present age laying the foundation for a new period of higher mental development, in the recognition and completion of the theory of descent. The recognition of the theory of development and the monistic philosophy based upon it, forms the best criterion for the degree of man's mental development.[171]

So Haeckel himself sat at the very top of his evolutionary tree. We might have known.

Only one more thing remains to be said. Haeckel's ultimate "proof" of all this lay in imperialism:

> The Mediterranean species, and within it the Indo-Germanic, have by means of the higher development of their brain surpassed all the other races and species in the struggle for life, and have already spread the net of their dominion over the whole globe.[172]

That was a dreadfully common European—and American—view at the end of the nineteenth century. And Darwin himself, alas, had come close to expressing it several times over in *The Descent of Man*: "The civilized races have extended, and are now everywhere extending their range, so as to take the place of the lower races."[173] But if that was Haeckel's "good news," how could Lu Xun accept it?

6

The History of Mankind

How could Lu Xun proclaim a gospel that led to racism and imperialism, when anti-imperialism was still his major theme? How could Lu Xun hail an apostle of mechanical materialism, who denied free will, when it was the will of his people that he wanted to awaken?

There is "a most ingenious paradox" in Lu Xun's introduction of the thought of Haeckel, that brings to mind the "enormous paradox" that Benjamin Schwartz saw in Yan Fu's introduction of the thought of Huxley. Professor Schwartz, believing Yan Fu to have embraced the Social Darwinism of Herbert Spencer, found it puzzling that Yan Fu should translate *Evolution and Ethics*, a work that was "decidedly not an exposition of Social Darwinism" but "actually represented an attack on Social Darwinism." Why, he wondered, should Yan Fu "choose to translate a work so little in tune with his basic message?"[1]

I have argued elsewhere that Yan Fu's choice was paradoxically not so paradoxical. "Correcting," in his commentary, Huxley's pessimism with Spencer's faith in progress, and Spencer's do-nothing faith in evolution with Huxley's call for noble action, Yan Fu ended up with the very fusion, through confusion, of determinism and determinationism that he wanted—a fusion and confusion of tremendous psychological, if illogical, fitness.[2]

But Professor Schwartz's question is well asked of Lu Xun, although Lu Xun chose to introduce Haeckel in an article not a translation, for Haeckel's racism, imperialism, and mechanical materialism seem indeed "little in tune" with Lu Xun's basic message. We must ask, therefore, what Lu Xun liked in Haeckel's works, and how he managed to overlook what he did not like.

Lu Xun introduced Haeckel to reintroduce the theory of evolution, clearly because he thought Yan Fu's introduction, ten years earlier, had failed to take.

That was a mildly odd conclusion, for Yan Fu's introduction had taken wildly well, not among the masses, who were illiterate, and not among high officials, who perceived, cleverly enough, in Darwin's theory, a threat to their dynasty, their traditions, and their rice bowls, but certainly among angry intellectuals like Lu Xun, among the droves of dissident patriots out of power who insisted that China's only hope of survival lay in change. They had taken to Darwinism, as they understood it, very easily. Liang Qichao had been waving Darwin's banner for a decade, in his impassioned arguments for reform and revolution. So had the Republican Sun Yatsen, so had the rabidly anti-Manchu anarchist, Wu Zhihui, so had hundreds of lesser polemicists. All of them, in their arguments, had turned to Darwin.[3]

So it was again an indication of Lu Xun's political loneliness that in 1907 he should still give no indication that he thought any of those arguments amounted to much. He echoed none of them, except for the earliest argument of Yan Fu—that evolution was for change and against conservatism (a moot point). "Recently in China," said Lu Xun, "the term evolution has almost become commonplace. Those who rejoice in the new, use it to embellish their rhetoric. Those loyal to the old, decry its classification of men with monkeys and resist it with all their might." It was not, of course, surprising that "conservative Chinese, clinging to their tattered traditions, should flee at the sound of something new." Even in Germany, "the center of learning," there were people, famous people, who opposed Darwin's theory of man's origins. The philosopher Paulsen had actually said, unashamedly: "It is to Germany's shame that so many people read Haeckel's works." But Paulsen, for all his learning, was a fool. Haeckel's works "were the very peak of modern biology." And Chinese needed to know it.[4]

Lu Xun chose Haeckel because Haeckel had written the latest word, Lu Xun thought, on evolution. Haeckel had surpassed the master: "Germany's Haeckel, like Huxley, has been a modern champion of Darwin's doctrine. But he has not just been faithful to the old [the sin of China's conservatives]. He has greatly improved upon it, making a systematic chart of the evolution of living things . . . from ancient single cells all the way to modern man. . . . Although scholars of later ages may make even further progress, *ad infinitum*, of those who discussed evolution at the end of the nineteenth century, surely this man's accomplishment was the greatest."[5]

It was no mean accomplishment for Lu Xun to summarize that accomplishment. That much we must admit whether we find Haeckel's accomplishment great or not. In 1907, Lu Xun's "History of Man," however brief, was, to the best of my knowledge, the best yet written in Chinese. It was a report, of course, not a paper. The only original things in it were a few unargued assertions. Nonetheless to have understood Haeckel, through either German or Japanese, to have come to terms with a host of terms, scientific and philosophical, that must still have been depressingly foreign, to have pieced together a very ill charted (in Chinese) tract of Western intellectual history, and to have reproduced it all in doubly erudite language was, indeed, an accomplishment. And when it was accomplished, Lu Xun perhaps knew more about evolution than Yan Fu.

But in displaying his new knowledge, Lu Xun also displayed his ignorance. He displayed his ignorance by accepting almost everything Haeckel said—uncritically.

He accepted uncritically, for example, Haeckel's evaluation of Linné, or Linnaeus. He praised Linné, for his system of classifying and naming species, pitied him for his adherence to the creation story of *Genesis*, and then praised him again for his ability to capture the natural traits of a species in a few deft strokes: "He recorded the special characteristics of each, so one could grasp them at a glance."[6] One can only hope that Lu Xun was ignorant of Linné's description of "*homo sapiens asiaticus*":

> Silky yellow, melancholic, stiff.
> Hair: nearly black. Eyes: dark.
> Severe, haughty, avaricious.
> Clothed in loose garments.
> Ruled by opinions.[7]

Lu Xun himself would later harshly analyze his people's "*guomin xing*, their "national character" or "nature." But here he showed no understanding of how silly it was for Linné, to lump together as universal racial traits, elements of appearance, temperament, and culture. He did not see that Linné's earnest efforts were laughable—but no joke. He did not see that Linné, who first brought order to Adam's earliest work, could not himself disclaim the name of godfather of most nineteenth-century racists—at least of "scientific" racists.

Lu Xun also uncritically accepted Haeckel's high opinion of Lamarck. Modern Chinese scholars have given Lu Xun great credit for independently recognizing Lamarck as a great, unrecognized, materialist predecessor of Darwin,[8] but Lu Xun was merely echoing Haeckel (although, granted, with approval), just as he was when he accepted Lamarck's belief in the evolutionary importance of use and disuse and in the inheritability of acquired characteristics.[9]

Lu Xun accepted uncritically Haeckel's biogenic law. But he could not have understood it very well, for the second time he mentioned it he got it backwards: "Phylogeny is a repetition of ontogeny."[10] One must assume that was just a slip, but Lu Xun never caught it—(nor in the last eighty years have any of his editors). In any case, he never showed the slightest awareness of the dubiousness of Haeckel's law—even when he got it straight.

Lu Xun accepted uncritically Haeckel's chart of evolution. He first accepted a very simplistic description of how species had evolved from one another. "Once the Cenolithic or tertiary epoch was reached," he said, "semi-apes appeared. Next were born true apes, among whom was a race of narrow-nosed apes, and from that race were born baboons and then man-apes. And man-apes gave birth to ape-men, who could not speak. But later they could speak. It is they we call men."[11] But when Lu Xun put this simplistic history of the descent of man into a phylogenetic chart (or when he reproduced what he deemed a fit one from some Japanese translation), low and behold it was a chart of man's *ascent* (as usual). It was diagonal not vertical, but Lu Xun's graph of man's ascent rose even more inexorably than Haeckel's tree.

Lu Xun's chart actually proved, of course, nothing but mankind's youth, but Lu Xun insisted it proved man's primacy, which was just what Haeckel thought his tree proved, man being, as we remember, evolution's "youngest and most perfect twig" of its "most perfect and most highly developed branch"—the order of the most aptly called "primates" (Linné's prejudice). That was Haeckel's dogma despite his own warning against "the prevalent illusion of man's supreme importance."[12]

Lu Xun too ignored that warning, assuming what Haeckel did, that it was only among the religious that one encountered "this boundless presumption of conceited man."[13] Failing to recognize his or Haeckel's own conceit, Lu Xun blithely assured his readers that "the theory of mankind's evolution has in no way challenged man's primacy. Man has progressed every day without stopping, from the lowly to the high. It is in this that we can see that in ability man surpasses all creatures. Why be ashamed of our origins?"[14]

That was Haeckel's argument.[15] It was Huxley's (except the part about steady progress).[16] It was even Darwin's.[17] But I hope it is clear by now that that argument was *not* scientifically Darwinian.

Lu Xun accepted uncritically Haeckel's equation of evolution and progress. But that was not just thanks to Haeckel's powers of persuasion. In accepting the equation of evolution and progress, Lu Xun was of Haeckel's persuasion before he ever heard of Haeckel. That was undoubtedly one of the reasons he liked Haeckel. For he had accepted the equation of evolution and progress when he first read *Tianyan lun*, where Yan Fu, quoting Spencer, introduced it in his main disagreement with Huxley.[18]

Almost all Chinese who read *Tianyan lun* assumed that the theory of evolution was proof of progress—and not just because Spencer and Yan Fu said so. It was so obviously the most satisfying conclusion to which to leap. Darwin's theory in China gave rise at once to two rival visions, one frightening, of countries and races going under, and one inspiring, of Utopia in the making. It was only "natural" that Chinese, in their "most dangerous hour"[19] (or through a whole series of them) should warn of the one and take comfort in the other.

Granted, Westerners confused progress and evolution before them. That confusion was the commonest confusion of the nineteenth century. But in China it became complete, for in China, in Chinese, the *two* words, progress and evolution, became one.

Yan Fu's word for evolution, *tianyan* (natural evolution), quickly proved less fit than his word for progress, *jinhua* (progressive change). So his term for the theory of evolution, *tianyan lun*, turned, overnight into *jinhua lun* (the theory of progressive change)—which is what Chinese call evolution to this day.[20]

That is why it is so difficult to talk about evolution in Chinese (*ming bu zheng ze yan bu shun*).[21] A Chinese scholar, writing, a hundred years after the birth of Lu Xun, on the theory of evolution in Lu Xun's early thought, says, "'A glance at the term, and its meaning leaps to mind [*gu ming si yi*]': '*Jinhua*' means change, evolution, development and progress."[22] Ecce the great confusion, the persistent fallacy. On the face of it, evolution does *not* mean progress. *In reality* it may mean progress, but if it does, there is more to both the name and the reality than meets the eye. In any case, to *prove* that the two are one, one must first see the one as two. So someday, Chinese must admit that *jinhua lun* is a mistranslation.

That mistranslation was not Lu Xun's. But he accepted it, never questioned it, and so helped perpetuate it. (And as a "sage of New China," he perpetuates it still).[23] Never asking how evolution could mean progress, he echoed the great confusion of the day: "Mankind made daily progress without end." "The great law of progressive change advanced animals to higher levels." "Natural selection . . . brought living organisms to perfect fitness." "The fact of mankind's progressive evolution is clear beyond the shadow of a doubt." "[Life] has gradually progressed from monera to man."[24]

Without realizing what he was doing, seemingly without thinking about what he was doing, Lu Xun slipped a teleological argument into his explanation of evolution, just as Haeckel did—although Haeckel, in all his works, repeatedly ridiculed the teleological arguments of others. Haeckel repeatedly ridiculed "the prevailing Doctrine of Design, or Teleology," which "the Doctrine of Evolution," he insisted, had demolished.[25] Darwin, in one fell swoop, had driven the designer off the stage of natural history and of human history (all the world being only one stage). He had driven him out of the theater: The set had no designer—the play had no playwright. Darwin had not merely done in the doctrine "that each species has been independently created."[26] He had shown that species had in no wise been created. He had disproved forever and ever, amen, said Haeckel, "the teleological view of nature, which explains the phenomena of the organic world by the action of a personal creator acting for a definite purpose."[27]

Lu Xun echoed that conviction. With obvious satisfaction he translated (and quoted in German) Haeckel's famous taunt that the popes, ultimate defenders of "the Doctrine of Design," were "the greatest charlatans in world history."[28] And then he smugly asserted as his own what was Haeckel's "self-evident" conclusion: "The theory of evolution was established upon refutation of the theory of divine creation."[29] But heralding that "refutation of the theory of divine creation," ridiculing the Western notion of a "personal creator acting for a definite purpose," Lu Xun, like Haeckel himself, still left his readers with a "teleological view of nature"— for he endowed evolution with direction.

"Evolution is like a flying arrow,"[30] said Lu Xun in a hastily drawn simile that still flies today, however much off target. In essay after essay Lu Xun's arrow simile has been cited by Lu Xun scholars in the People's Republic to prove Lu Xun's prescient recognition of the inevitability of "The Revolution,"[31] and to that end, if only to that end, it has been cited with some, if only some, reason. In 1907, Lu Xun, of course, as yet had not the foggiest notion of Communist revolution, but he did imply, though he did not yet say so outright, that "evolution" depended on revolution, that progressive change *(jinhua)* inevitably came through violence. "'Peace,'" he said, in the least peaceful paragraph he ever wrote, "does not exist on this earth. What is forcedly called 'peace' is only a superficial quiet after war, or before it." For all the while "fierce fires rage underground" (a proto-Marxian metaphor), just waiting to erupt.[32] And erupt they shall— bringing progress.

It was then that he fired his arrow (Chinese metaphors mix easily) at "reactionaries" (although he did not yet have a name for them, Progress having to be established before reactionaries could be). He fired his arrow at China's sages who had resisted Progress in the name of peace. That was Yan Fu's argument of twelve years earlier: China's sages, Confucian, Daoist, Buddhist, by idealizing peace and condemning struggle, had hindered China's progress for two millennia, and they did so still.[33] "But unfortunately evolution is like a flying arrow," said Lu Xun. It will not stop until it falls. It will not stop until it hits. You may pray that it fly backwards, back to the bowstring, but there is no such physics."[34]

What did that mean? Whom was he really attacking? Politically it was perfectly clear. He was shooting at Chinese conservatives. All he really was saying was "Oh, you benighted Chinese! Stop resisting reform.

There is no returning to 'the good old days,' no preserving 'the good old ways.' China has got to change!"—And who would not shout, "Amen?" But again, one has to distinguish his rhyme from his reason. Lu Xun was echoing Yan Fu and Liang Qichao.[35] He was echoing the positive side of their "Darwinian" warning, "You can't stop progress," but that warning was *not* Darwinian because unfortunately, *evolution* was *not* like a flying arrow. Lu Xun's simile made no sense.

He meant it to mean two things, but it was proof of neither. He meant it first to say that people cannot revert to older ways because biological evolution allows no devolution, no retrogression. (He forgot his earlier threat that Chinese, if they did not watch out, might "degenerate day by day becoming apes, then birds, then clams, then algae, and finally inorganic things."[36]) Arrows cannot return to their bow strings, "living things cannot return to their origins,"[37] and we cannot return to, or persevere in, the ways of the ancients.

But that does not follow. The improbability of the inhabitants of Peking (if I may return, for a moment, to the spelling of the ancients) ever having descendents resembling Peking Man does not prove that they could not have an emperor if they wanted one. It would take a Monist of Haeckelian mettle indeed to declare without flinching that one simple law governs arrows, organisms, and organizers. Edward O. Wilson himself could not shoot an arrow as easily as Lu Xun did, across the chasm (though somehow it must have a floor) between human evolution and human history. Lu Xun shot without thinking. He was not of Haeckel's mettle. He was not trying desperately to understand evolution. He just knew what he did not like.

But even on biology's uncontested side of the chasm, it is not easy to find arrow heads. Devolution is not strictly speaking impossible. It is just *extremely* unlikely. The overworked Birmingham moths have proven that evolutionary trends may be reversed. The only reason it would be *virtually* impossible for an organism to "return to its origins" is that it, or its descendents, could only get there by going around a million corners, corners rounded the first time only by an incredible series of coincidental mutations and environmental changes. To get back around those corners by a reverse series of coincidences would require a coincidental tour-de-force to end all coincidental tours-de-force. So devolution over any distance is unlikely—because of the *unarrowlikeness* of evolution.

Actually, Lu Xun granted the unarrowlikeness of evolution, in a sense, in a second essay, in his "Lessons from the History of Science." Surveying the evolution or progress of science, he said, "Surely it is a true saying that the world's progress is not straight but often twisted like a spiral."[38] That saying, already a commonplace (who knows who said it first), already (although Lu Xun did not yet know it) a tenet of Marxism, one destined to be enshrined in the thought of Mao Zedong ("the future is bright, but the road is winding"[39]) put Lu Xun, metaphorically, in trouble: the mythical archer Hou Yi could not have shot arrows that wound.[40] But philosophically Lu Xun was in even greater trouble, for however incompatible his metaphors, he ultimately meant them to mean one thing, the second thing he had meant his arrow to mean. Whether the arrow of evolution flew straight or somehow ricocheted down a corridor of space and time, its direction was set. Evolution was not aimless. It was aimed. And it would not stop until it fell to earth, until it hit something. But how could it be aimed? Lu Xun categorically denied the existence of an archer, all the while that he insisted evolution was like an arrow. As a would be Monist, he shot himself in the foot.

Lu Xun, in his early essays, and beyond, professed to have faith in inevitable progress. That was Yan Fu's faith, borrowed from Spencer, declared in *Tianyan lun*, in the very teeth of Huxley's arguments against it. In his concluding comment to Huxley's "Prolegomena," Yan Fu said, "What we can know today is that the world must progress and that the future will surpass the present."[41] Spencer was the authority: "Spencer has said that society, as long as it evolves naturally, must make daily progress towards good and away from evil, and perfect order must someday come. His argument is extremely solid. Indeed it is well nigh indestructible." Why? Because his social principles were derived from biological principles, and it was now clear that "evolution causes living things to make daily progress." Darwin confirmed that faith.[42] Haeckel more than kept it. So Lu Xun now embraced it: "The fact of the progressive evolution of mankind is clear beyond the shadow of a doubt."[43]

But actually, from beginning to end Lu Xun was racked with doubt, because whatever he thought of evolution, he had a very low opinion of humanity. Even in 1908 he could, on alternate Tuesdays, sound Huxleyian: "For the disappearance of warfare and everlasting peace, we will have to wait for the extinction of man and the dissolution of the earth; so arms and armor will be as long lived as the human race."[44] And on a dark Huxleyian day in 1924, he could virtually renounce his evolutionary faith: "I do not believe very much in historical progress. . . . Our ancient nature will always show itself."[45] And yet in 1928, in his most famous pronouncement on the subject, he professed to have kept the faith at least until 1927, when he witnessed, to his great disillusionment, young Nationalists slaughtering young Communists (what happened to his faith thereafter, we shall have to consider anon). He declared, almost quoting from Yan Fu, "I always believed in evolution, I always thought that the future must surpass the past and that the young must surpass the old."[46]

Lu Xun believed what Chairman Mao said, decades before Chairman Mao said it: "The world will go towards progress; it is definitely not going towards reaction,"[47] because Lu Xun believed, in one more rather troublesome metaphor of his own, that "the devil . . . cannot cover up the light."[48]

Why did all this leave Lu Xun in philosophical trouble?—Because faith in inevitable progress is a metaphysical faith, and Lu Xun was introducing Haeckel's Monist materialism.

Lu Xun accepted Haeckel's Monist materialism—uncritically. He did not wonder how matter could have meaning. He did not wonder how matter could measure matter. He did not wonder how matter could be mind. He did not ask philosophical questions. Man was matter and that was that: "The entire material world was formed through cause and effect. The phenomena of the universe all obey that law. Thus if we seek for the origin of organic matter, which is made of inorganic matter and which will in the end revert to inorganic matter, it must be inorganic matter. . . . That the inanimate becomes animate has become an irrefutable truth. Behold how startling is the science of the late nineteenth century. As to the origin of inanimate things, we must await word from cosmogony."[49]

That is how he ended his "History of Man"—with an echo of the master know-it-all: The last riddle of the universe would soon enough be solved.

So Lu Xun also accepted, uncritically, Haeckel's know-it-all-ism: "The progressive evolution of mankind is clear beyond the shadow of a doubt."[50] Darwin "swept away all doubts."[51] "All doubts melted away."[52] "There was nothing that could not be explained by natural law."[53] "The great mystery was clear."[54]

Haeckel's optimism, Haeckel's hubris. But why quarrel with Lu Xun for accepting it? Optimism came with such difficulty to Lu Xun that it seems unkind to criticize him on those rare days when he managed to show some. Nonetheless, in echoing Haeckel's optimism and Haeckel's hubris, however innocently, however understandably, he did, I think, his people a disservice, not a great disservice, because his early difficult articles were not very influential, but still Lu Xun helped, unwittingly, propagate a superstitious faith in science, especially "social science," that would eventually help people "eat people." Even Lu Xun helped Yan Fu, and Liang Qichao, and Wu Zhihui, and Sun Yatsen, and Hu Shi, and Chen Duxiu and untold others spread "Scientism," as Daniel Kwok called it years ago, "Scientism," the unnamed first new faith of those who lost faith in "the Great Tradition."

Given the dreadful uncertainty of China's fate and the uncertainty of Chinese patriots about what to do about it, the attraction of Haeckelian certainty about science and social progress was certainly understandable.

The promise of "Science," a *Dao* above East and West, open to all who would follow it, a saving force for all in the know, was overwhelmingly attractive.

But the hubris in Haeckel's faith that the elect, at least, could know all, could and would lead to a worse kind of hubris, to one close to the original meaning of the word. Vain pride could turn into arrogant violence. Know-it-alls, convinced that they and they alone knew the science of social evolution, and that, armed with that knowledge, they and they alone represented the Forces of History, could convince themselves that they had a natural right to rule, a scientific right as absolute as any divine right of any potentate of old. And so—pathetic irony—the socially scientific could show a religious self-righteousness, a raging intolerance, and, although this should have been contradictory, a murderous insecurity to equal that of the Inquisition: "All demons and devils shall be annihilated."[55]

Chinese Marxist historians, however, have seen in Lu Xun's acceptance of Haeckel's monist scientism proof only of an admirably early bent toward materialism. From his view of nature one could see that even in his early thought "philosophically he resolutely held to the path of militant materialism," a materialism "based on Darwin's theory of evolution."[56] And in one sense that was true. Lu Xun wanted to be a materialist, as Haeckel did. But neither Haeckel nor Lu Xun faced, nor have Chinese Marxist historians, so eager to welcome Lu Xun at the earliest possible moment into the fold (although they disagree as to when that moment was) ever faced, the contradiction between monism and a teleological faith in progress. The irony, however, this time not grim but amusing, is that the pseudo-materialism of Haeckel and Lu Xun was precisely the kind of materialism necessary for belief in Marxism, itself metaphysical to the core. So in that sense, Lu Xun was, indeed, preparing the way for the thought of Mao Zedong.

Lu Xun did not accept Haeckel's ranking of races or his social justification of imperialism. How could he—without accepting for his people an inevitable, natural bondage? But how did he get around Haeckel's views? He did not specifically refute them. He did not specifically acknowledge them. Was he ignorant of them or did he deliberately ignore them?

It would be easy enough to imagine that Lu Xun simply missed Haeckel at his worst, that he never read, in either German or Japanese, the racial chapters of the second volume of *The History of Creation*. But one thing blocks that easy way out of our paradox: Lu Xun used one racial argument that did, in a twisted way, come straight from Haeckel. He used one peculiar argument peculiar to Haeckel's argument.

Now if that proves that Lu Xun knew of Haeckel's racial views (we shall get to the argument itself in a moment), we are forced to a pair of paradoxical conclusions. First, Lu Xun dared to pick and choose. He felt no obligation to accept Haeckel's thought in its entirety. It seems to me that he never accepted anyone's thought in its entirety. Like Yan Fu before him, he never acknowledged any Western wise man as his master. To the end he displayed a critical independence that, had he lived long enough, would have gotten him in trouble, probably in 1942, certainly after 1949.

In this case, however, Lu Xun did not dare, or he did not choose, to attack directly those of Haeckel's views that he did not accept. He did not dare challenge, directly, Haeckel's theory of racial inequality. Why not? Because the whole question of racial equality or inequality was too dangerous. It hung over the heads of all Chinese, because Westerners, all kinds of Westerners, merchants, soldiers, scientists, priests, said Chinese were unfit. And the White Peril threatened to prove it. Until one could *disprove* it, the question was better left unraised.

I do not think Lu Xun thought his people unfit. At least he did not think them biologically unfit. He thought them benighted, and he did fear the White Peril, but I think we shall see that even when he most despaired of his people standing up to the White Peril, he never gave up absolutely the faith Mencius showed before King Xuan. He never feared that his people *could* not save themselves. He feared that they *would* not.[57] He was himself to become his people's harshest critic, but he would not give in to the Social Darwinian contention that his race was evolutionary lowlife.

Why then did he not challenge Haeckel directly? Because he did not know how to challenge him on his own ground. He could not fight his way through Haeckel's "facts," because he was not a scientist. Nonetheless he did attack him indirectly. Without naming him, he hoisted Haeckel with his own petard. With poetic justice, if no other, he turned the fallacious argument with which Haeckel justified imperialism into an equally fallacious argument *against* imperialism.

In doing that, he neatly gave Social Darwinian, racist imperialists a bit of their own—which surely was no less than they deserved. But a bad argument was a bad argument even in a good cause. Had Lu Xun been being facetious, or wittingly sophistical, we could smile and go on, but he was not. He had not yet learned to wield, with any deftness, the "daggers and spears"[58] of satire that would so justly make him famous. Later he would sometimes choose to ridicule the ridiculous with ridiculous arguments,[59] but in his early essays, for the most part, he still "meant what he said, and he said what he meant."[60] So his anti-Haeckelian Haeckelian argument is revealing.

Scholars in the People's Republic have never questioned that argument. They have been happy enough to cite it as proof that Lu Xun opposed imperialism. But we already know that Lu Xun opposed imperialism. What we must determine is *why* he opposed imperialism. On what philosophical grounds did he oppose imperialism?

In his anti-Haeckelian argument against imperialism Lu Xun revealed two unacknowledged theories, a theory of human nature and a theory of ethics—a theory of evolution and ethics. We must work them out. Lu Xun never worked them out for us. He never worked them out for himself. And yet they lie at the very heart of his early thought. They lie at the very heart of his thought early, middle and late—they lie at the very heart of his thought, period, inextricably entangled, fraught with contradictions, each a part of the other and yet each, in part, at odds with the other—and each at odds with his materialism. Together they form the last great knot that we must try to unravel if we are to come to any understanding of Lu Xun's thought.

But expect no blade to cut that knot asunder. At the heart of Lu Xun's thought there lies an unresolved contradiction.

7

On Human Nature

The anti-Haeckelian anti-imperialist argument in which Lu Xun revealed a theory of human nature and a theory of ethics was his famous argument against "brutish patriotism," or, more literally, against "animal-natured patriotism" (*shou xing zhi ai guo*).[1]

"Animal-natured patriots" were imperialists. Lu Xun still did not use the new Chinese-Japanese word for "imperialists" (*diguozhuyizhe*), then recently made in Japan, but the species he described was unmistakable, even if his description was mildly exaggerated. "Animal-natured patriots" were "aggressors," beings "addicted to killing and conquering, bent on extending their nation's sway over all the world. . . . Animal-natured patriots . . . respect no nation but their own. They look down on foreign lands. Seizing on the 'survival of the fittest' language of evolution, they attack the weak to indulge their desires, and they shall not be satisfied until they have forced the world into one and made all alien races their vassals."[2]

Now there was nothing odd in that assessment. Chinese had good reason to think poorly of imperialists. What counts is that Lu Xun, in calling imperialists brutes, was not just indulging in invective. He was himself "seizing upon the 'survival of the fittest' language of evolution" to explain the existence of imperialists—but in a way exactly opposite to the way in which they explained their own.

Imperialists were evolutionary throwbacks. Despite their seeming fitness in the present world, they were subhuman, backward elements, "reactionary" elements, who were actually on their way out. As the Chairman would say, much later, they had a "double nature." They were in the present "real tigers, that eat people," but they would, "in the end, turn into paper tigers, dead tigers, bean curd tigers."[3] *Nota bene:* this ahistorical usefulness of Chairman Mao's language is not coincidental. Lu Xun was helping, with his own language, partially borrowed, of course from others, to prepare a foundation for the thought of Mao Zedong. Lu Xun clearly believed in an idealistic chain of evolutionary being, in which imperialists appeared as imperfect human beings. "Human beings," he said, "from their origin as microorganisms, from worms to tigers and leopards and apes, unto the present day, have had an ancient nature latent within, which time and again has revealed itself. That is why such things as the love of killing and aggression exist, and the seizing of others' territory, and sons and daughters, and jade and silk, to satiate wild greed."[4]

Had Lu Xun written no more than that, we could more readily hear echos of Huxley and Nietzsche than any echos of Haeckel.

We can clearly hear Huxley. The apes and tigers were his. So was that "ancient nature latent within." "Ape and tiger promptings" were "instincts of savagery," of "unlimited self-assertion," that were "men's inheritance from the ancestors who fought a good fight in the state of nature," "their inheritance (the reality at the bottom of the doctrine of original sin) from the long series of ancestors, human and semi-human and brutal, in whom the strength of this innate tendency to self-assertion was the condition of victory in the struggle for existence." That inheritance, "the cosmic nature born with us," our "dose of original sin,"[5] was the continuing cause of our "inhumanity"—of which imperialism was but one manifestation.

And we can hear Nietzsche, whom Lu Xun had just discovered—again in translation.[6] The "worms" were Nietzsche's (as were half of the apes). Human beings, said Lu Xun "got their start as microorganisms" and then evolved "from worms to tigers and leopards, to apes"—to us. But *"some have kept the nature of a worm, and some the nature of an ape."*[7] Thus spoke Lu Xun. Thus spoke Zarathustra: "You have made your way from worm to man, and much in you is still worm. Once you were apes, and even now man is more of an ape than any ape."[8]

That sentence meant so much to Lu Xun that he would still see fit to cite it twenty-eight years later in 1935, the year before he died.[9] But whatever it meant to him then, when first he echoed it in 1907, his echo was imperfect. In one all important respect Lu Xun did not mean by the ape and tiger or worm and ape in mankind exactly what either Huxley or Nietzsche meant. For Huxley saw the ape and tiger in all of us,[10] and Nietzsche saw the worm and ape in most of us—in "man." Lu Xun saw worm and ape and tiger in *imperialists*, and "feral patriots."

In two all important sentences, we hear Haeckel. Lu Xun said we must "take into account the *nonuniformity* of the human race" (*cha renlei zhi bu qi*). For he said, "There are great differences in the degree to which men have walked the road of evolution. *Some* have kept the nature of worms and *some* the nature of apes. So even in ten thousand years there can be no 'great unity.'"[11]

That is what Haeckel said of *races*. *Races* represented "different degrees of perfection" along "the road of evolution." In *races* one saw "the nonuniformity of the human race." And in imperialism one saw the natural result of that nonuniformity: "the dominion" of the fit races over the less fit.[12]

Lu Xun turned that argument upside down. The "feral nature" of imperialists proved their low level of evolution. Their fitness was illusory. Their dominion would be short lived. Their future was extinction. They would perish from the face of the earth—someday.

Lu Xun did not, however, identify his imperialists with the white race. Haeckel's imperialists were the whitest of the white, members of "the Mediterranean species and within it [of] the Indo Germanic."[13] Lu Xun's imperialists could come in any shade. They distinguished themselves, they revealed their "feral nature," not in their appearance, but in their behavior. So Lu Xun seemed to rise above Haeckel's racism. And yet his imperialists were still a breed apart, a biological breed apart. His explanation of imperialism was as "naturalistic" as Haeckel's, and as "Darwinian" (and as unDarwinian)—because it was based on Haeckel's evolutionary theory of human inequality.

Lu Xun used Haeckel's theory to attack Haeckel's theory. To use the rather useful rhetoric of the (much later) Cultural Revolution, he waved Haeckel's flag against Haeckel's flag. In denouncing those who "seized upon the 'survival of the fittest' language of evolution" to justify imperialism, he waved the flag of evolution against the flag of evolution.

Rather neatly, Lu Xun did to imperialists what Thorstein Veblen did to capitalists. Veblen, in *The Theory of the Leisure Class*, had eight years earlier, in 1899, called *capitalists* atavistic throwbacks. Capitalists too were only "real tigers" for a while, and so were really "bean curd tigers,"—because they owed their "fitness" to "predatory aptitudes and propensities carried over by heredity and tradition from the barbarian past of the race." They stood, therefore, in the way of evolution and would be run over by it. Veblen turned the weapons of Herbert Spencer and Cesare Lombroso (and Lombroso was a Haeckelian) against the class most practiced in the use of those weapons for offensive self-defense.[14] Lu Xun knew no more of Veblen than Veblen knew of Lu Xun, but both fought Social Darwinism with Social Darwinism.

But beware, all ye who are tempted to applaud, all who would like to believe that capitalists and imperialists are indeed *yi qiu zhi he* (raccoon dogs—to give apes and tigers a rest—from the same mound). Lu Xun's and Veblen's arguments attacking imperialists and capitalists were no more scientific than Haeckel's, Spencer's and Lombroso's arguments defending them. Darwin, Huxley, Haeckel, Nietzsche—none proved Lu Xun right, neither, in recent times, have any or all members of the Naked Ape school. Some might think that Lu Xun now has a troop of anthropologists on his side, but he has not.

In the 1950s and 1960s, such a troop did make much, once again, of "the ape within us." Raymond Dart argued in his essay, "The Predatory Transition from Ape to Man," that there was still a "killer ape" within us. "The toothsome cruelty of mankind to man," he said, "forms one of his inescapable characteristics and differentiative features; it is explicable only in terms of his carnivorous and cannibalistic origin."[15]

In *African Genesis*, Robert Ardrey called man "a predator whose natural instinct is to kill with a weapon." To that instinct, he said, we owed our evolution. That instinct had separated us, literally, from other apes. And that instinct was still with us.[16]

Then Konrad Lorenz came out with *On Aggression*, a study of the causes and effects of "the aggression drive," which he accepted without question as an evolutionarily endowed component of human nature.[17] Lorenz had found "unexpected correspondences" between his work and that of Sigmund Freud, who he thought had written more than enough on aggression decades before.[18] Freud had had a hard time making up his mind about whether man's "inclination to aggression" was or was not an independent drive (independent of the libido), but in *Civilization and Its Discontents*, he had finally declared "that the inclination to aggression is an original, self-subsisting instinctual disposition in man."[19] Lorenz set out to explain how that could be, entreating us "as good Darwinians" to "inquire into the species-preserving function" of "fights within the species." To his own satisfaction, he demonstrated quite easily the "survival value of intra-specific aggression."[20] (Huxley, after all, had stated seventy years earlier that the "innate tendency to self-assertion was the condition of victory in the struggle for existence."[21]) But Lorenz also made it abundantly clear that, valuable or not, "the aggression drive," as the Sage King Shun had said so long ago of human feelings in general, was "most unpretty."[22]

Huxley had said that "for his successful progress, throughout the savage state, man has been largely indebted to those qualities which he shares with the ape and the tiger," to "the instincts of savagery."[23]

Freud had said "cruel aggressiveness . . . reveals man as a savage beast to whom consideration towards his own kind is something alien."[24] It was in explaining the latter point that Lorenz made his contribution. He insisted that evolution had not endowed us with the "inhibitory mechanisms" that keep many other animals from slaughtering their own kind.[25] Freud paraphrased Plautus' observation of twenty-two hundred years ago: "*Homo homini lupus*" (man is a wolf to man),[26] but Lorenz insisted that man was more of a wolf than any wolf (again hear Nietzsche: "Man is more of an ape than any ape"), because wolves were *not* wolves to wolves, or rather *in being wolves to wolves*, wolves rarely *killed* each other, because their "intra-specific aggression" drive had been blunted by what were among "the most reliable [intra-specific] killing inhibitions in the world,"[27] killing inhibitions that we, alas, have not.

"Civilized man," therefore, is more brutal than any brute, and more savage than any savage.[28] Did that not vindicate Lu Xun? One must admit that in two ways Lorenz came very close, unwittingly, to echoing Lu Xun. He gave this portrait of an "animal-natured patriot" displaying "militant enthusiasm":

> The tone of the entire striated musculature is raised, the carriage is stiffened, the arms are raised from the sides and slightly rotated inward so that the elbows point outward. The head is proudly raised, the chin stuck out, and the facial muscles mime the "hero face," familiar from the films. On the back and along the outer surface of the arms the hair stands on end. This is the objectively observed aspect of the shiver!
>
> Anybody who has ever seen the corresponding behavior of the male chimpanzee defending his band or family with self-sacrificing courage will doubt the purely spiritual character of human enthusiasm.[29]

Moreover, Lorenz linked this apelike militant enthusiasm to imperialism. "Very large numbers inspired by the same enthusiasm," he said, would "feel the urge to conquer the whole world."[30]

Finally Desmond Morris dubbed man, in 1966, "the naked ape," a mere two thousand years or so after a Chinese scholar called man a "naked beast."[31] Morris was willing to grant man more apelike intraspecific killing inhibitions than Lorenz was, but sounding neo-Malthusian fears, he was not at all sanguine about those inhibitions being up to the task of controlling "in 260 years time . . . a seething mass of 400,000 million naked apes crowding the face of the earth,"—considering "how little, how very little, the naked ape has changed since his early primitive days."[32]

But has all this ape talk proved Lu Xun right? No, first of all because the naked ape troop has not yet proved itself right. The school still has its opponents, two of whom, Richard E. Leakey and Roger Lewin, have charged that the above protestations "have been woven together to form one of the most dangerously persuasive myths of our time: that mankind is incorrigibly belligerent; that war and violence are in our genes."[33] They sound like Kropotkin attacking, in *Mutual Aid*, the "atrocious article"[34] in which Huxley "represented primitive men as a sort of tigers or lions," among whom, in Huxley's own words, "the Hobbesian war of each against all was the normal state of existence."[35] (Freud too believed in "the hostility of each against all and of all against each"[36]). Be that as it may, the more important reason the naked ape school has not proven Lu Xun right is that it sees apes in all of us. Lu Xun, in 1908, saw the nature of apes (or "worms"!) in *imperialists*.

Working from Haeckel's dogma that races of human beings could represent different levels of evolution, Lu Xun said imperialists were *biologically* backward—whatever their hue. But no modern anthropologists have said that. Lenin himself never said that. And it is now fairly clear that it was never so. The sun has set on the British Empire with no obvious rise in the level of British genes.

Lu Xun's early theory of "animal-natured" imperialists was "racist"—unless, of course, the whole thing was nothing but nationalistic hyperbole.

I do not think it was, but in any case we must take Lu Xun's anti-Haeckelian Haeckelian anti-imperialist argument seriously because of the theories of human nature and ethics that came from it, theories *not* forced out of three sentences, but visible many times over in some of Lu Xun's most famous and influential short stories and essays, theories of deep theoretical and practical import—of deep philosophical and political import.

To see these theories clearly it will indeed help to look ahead, to Lu Xun's first vernacular short story, "Kuangren riji" (A madman's diary) and to his "meanest" essay, "Lun 'feie polai' yinggai huan xing" (On the advisability of postponing 'fair play'), the one published in 1918, the other in 1925. It will seem less ahistorical, I think, to do so, when we see that both works are firmly rooted in Lu Xun's early thought. In what follows, we must also keep in mind Lu Xun's second most famous "slogan" (after "Save the children," the last words of "A Madman's Diary"[37]): "The most important thing is to reform our people's nature" (or "our national character" *guomin xing*).[38] And we must keep an ear open, again, for Nietzsche.

One more caveat. Lu Xun was not a philosopher. He did not sit down to write a treatise on human nature. He was not consistent in his terminology, and not absolutely consistent in his argument. What he said in 1908 was not exactly what he said in 1918. Nonetheless, out of a bit of initial confusion, a theory of human nature does emerge.

In 1908, thirty-four years before Mao Zedong in his famous, and infamous, "Talks at the Yan'an Conference on Art and Literature," declared that there was no human nature, or at least that there was no "abstract human nature," or "supra-class human nature,"[39] and fifty-two years before the literary critic Ba Ren got in trouble in the People's Republic for perversely persisting, despite the Chairman's teaching, in his belief that there was "a universal human nature,"[40] Lu Xun denied the existence of a universal human nature with his evolutionary notion of "the nonunifor-mity of the human race," that is, he denied that human beings as yet had any universal nature. They had different natures, reflecting different levels of evolution. And yet Lu Xun at the same time clearly did believe in an "abstract human nature." "Human nature" (*renxing*) was what was com-ing. It was evolving. It was in the making. It was not the nature of most human beings of the present. It was the nature of the "true human be-ings" (*zhen de ren*—the term he used in "A Madman's Diary") of the future.[41]

"Mankind," Lu Xun said, "has not yet grown up."[42] Most men (and women, of course) were somewhere in between beasts and "true men." Some were more "human" than others; some were more "brutal." It was, in general, as Nietzsche said: "Man stands at the middle of his course be-tween animal and Superman." "Man is a rope fastened between animal and Superman."[43] But for Lu Xun "man" was not at one point in the mid-dle. "Men" were at many points. Nietzsche too, it is true, said that "men are not equal,"[44] but he seemed to be thinking, when he said that, of the few who stood above the crowd, of "the enlightened men" or indeed "the supermen" already struggling to "overcome man." But "man" was one species, one disgusting species, from which, Nietzsche said, "our way is upward, from the species across to the super species."[45] For Nietzsche, in-equality was the way up. When Lu Xun first spoke of "the nonuniformity of man," he was looking in the other direction at our differing degrees of persistent "pre-human" brutality.

Lu Xun was not as optimistic as Nietzsche. In 1908 he may have been relatively optimistic. In an exceptionally patriotic moment, he called his own people very nearly "human." The imperialists were beasts, but the Chinese "treasured peace, as few others on earth," and so resembled true men:

Hating to spill blood, hating to kill people, abhorring separation, finding contentment in labor—*human* nature is like that. If only the whole world behaved as China did, then what Tolstoy said could be true: All the races on this earth, all the different states, would respect each other's borders and not invade each other, and order would reign for ten thousand generations.[46]

But Lu Xun would rarely again call his people good natured. In 1918 he accused his people of "eating people." His "madman" saw the light. He saw the moonlight and "finally realized that for more than thirty years"—Lu Xun was thirty-seven—he "had been totally befuddled." But now he saw the truth:

All things must be studied, before you can get them straight. People often ate people in the past. I remembered that, but I wasn't very clear about it. So I opened a history book to look it up—this history had no periods—and found, scrawled over every page, the words *ren-yi daode* [benevolence, righteousness, and morality]. I couldn't get to sleep in any case, so I perused the book half the night—before I made out words between the words—two words that filled the book: *chi ren* [eat people].[47]

This was Lu Xun's great indictment—of his own people. No longer did he concentrate his attack on the "feral nature" of imperialists. The crowd (from that far off slide) that "uttered a wolf-like cry" as it watched Ah Q carted off to his execution was a Chinese crowd. The eyes that followed Ah Q, more wolflike than the eyes of the real wolf who once "followed him, neither too close nor too far, wanting to eat his flesh—those wolf eyes, both fierce and cowardly, shining like two demon fires,"[48] were Chinese eyes—and not just Chinese landlord eyes, or merchant eyes or official eyes. The Hon. Mr. Zhao ate people. The tenant farmers of Wolf Cub Village ate people.[49] Old Fourth Master Lu and his household ate Xianglin Sao. The "short coated" customers of the All's Well Tavern, "workers," ate Kong Yiji.[50] Even Ah Q, eaten by all, ate people—or bit at them. The strong ate the weak. The weak ate the weaker. The Chinese people ate people. The Chinese ate Chinese.

After the failure of the successful Revolution of 1911,[51] Lu Xun never again spent time attacking the "feral nature" of foreigners. He spent the rest of his life exposing, bemoaning, satirizing, and trying to change his own people's "national nature," which, though different, was as brutal and barbaric in his eyes as "the nature" of imperialists. In "A Madman's Diary" he said his own people were inhumane and thus inhuman, and he explained that charge in the same "evolutionary" language with which he had first put white "brutal patriots" in their place:

Brother, probably in the beginning all wild men ate a little human flesh. Then later, because their thinking changed, some of them stopped eating people. They were determined to become good and so became men, true men. But some still ate people, and like insects some became fish and some birds and some monkeys—all the way to men, while some, who did not want to improve, stayed insects to this day. How ashamed should people who eat people be before people who don't eat people. I'm afraid they should be much more ashamed than insects before monkeys.[52]

Now this was the ranting of a "madman." We are under even less obligation to take it literally than we are Lu Xun's own talk in his essays of 1908, and yet the basic Nietzschean scheme of "A Madman's Diary," the notion that "true man" or "superman" would (somehow) evolve out of "man," or whatever man-beast we were, was a notion that Lu Xun did not give up. In 1919 he wrote:

> Although a Nietzschean type superman is too remote, we can be confident, if we judge from the reality of the human race in the world today, that someday in the future a higher and more perfect human species shall appear—at which time, I am afraid that above the man-like apes will have to be added the term ape-like men.[53]

Apelike men were what we were. With Nietzsche Lu Xun believed that "enlightened men" should recognize the truth (*pace* Oliver Hazard Perry and Walt Kelly), that we have met the missing link and we are it.[54]

Lu Xun's "madman" was neither "true man" nor "superman." He was only an "enlightened man"—but he was Lu Xun's hope. In 1935, the year before he died, Lu Xun still stood up for his madman, for he was not, he said, "as remote as Nietzsche's superman."[55] The madman was a link to the superman, one who would come before him, a true prophet in the wilderness, who would both condemn the present and promise a better future—urging eaters of men to repent.

Lu Xun's madman believed what John Stuart Mill believed, that "even the men and women who at present inhabit the more civilized parts of the world. . . assuredly are but starved specimens of what nature can and will produce."[56] A "higher and more perfect human species" would appear. That was the promise. But there was also a threat: "The future will have no place in the world for people who eat people." "If you do not change," cried the madman, "you yourselves shall all be devoured. However fertile you are, the true men will wipe you out, as hunters wipe out wolves—like insects."[57] So, very much indeed like a biblical prophet, Lu Xun's madman, still echoing Nietzsche—"Man is something that should be overcome"[58]—called on people eaters to repent: "You can change. Change from the bottom of your hearts!"[59]

Lu Xun's madman said Confucian morality ate people. But the madman was a Confucian prophet. He was one of Mencius' *xian zhi xian jue,* one of "the prescient," who "knew first and realized first" and so could awaken those who realized things more slowly.[60]

Lu Xun echoed Mencius' language many times over. The awakener was his hero, "the man who was awake"[61] who would awaken others. He was the "individual" Lu Xun wanted Chinese to "esteem."[62] He was "the genius," Lu Xun wanted Chinese not to "kill"[63]—or he was the forerunner of "the genius," for sometimes "the genius" was "the superman."[64] At any rate the awakener was the awakened man, the enlightened man, the only hope "until there is a genius:[65] "Only when the superman appears will the world have peace. If that cannot yet be, then all depends on the wise man."[66]

The "madman" was the wise man, the man who had eyes to see and ears to hear, who could prepare the way for the superman, for true man, who could make straight the way by condemning man's inhumanity to man.

The only trouble was, the madman "recovered." That was the perverse stroke of "genius" in Lu Xun's story. Lu Xun ended his story with a cry of hope that was forlorn enough:

> Maybe there are still children who have not eaten people. Save the children.[67]

But he had already thrown that doubtful hope into even greater doubt in his first paragraph—before we could understand it—by telling us that his madman, who would plead with us to overcome ourselves, had himself surrendered. He had "recovered," he had rejoined society, he was an official awaiting appointment—awaiting an opportunity to eat people.[68]

But what had he surrendered to? His nature? Just before his final cry he said, "I, with a four thousand year old history of eating people, although I did not know it at first, I now understand, I will have a hard time seeing true men."[69]

Despite Lu Xun's protestations of belief in the evolution of true men, he needed help with his disbelief. In 1918, before he wrote "A Madman's Diary," he wondered whether it was worthwhile waking people who were sealed in an "iron room." In the end, he wrote his story only because he could not prove his people had no hope.[70] In 1908, however, he himself had said he had great hope:

> I have not given up my great hope for the future. So I listen for the true voice of the knowing, and look for his inner light. Inner light is what will break the darkness. A true voice will eschew deceit. For a people, those two things will be like thunder in the early spring. A hundred grasses will begin to sprout, the color of dawn shall light the east, and the dark night shall pass.[71]

Still, in 1908 he had also almost voiced despair: "The sound of the first awakened comes not to break China's desolation."[72] We have already seen that he could sound like Huxley, in 1924: "I do not believe very much in history's progressive evolution. . . . Our ancient nature will always show itself";[73] and like Yan Fu, in 1932, "I have always believed in progressive evolution. I always thought the future would surpass the past, and the young would surpass the old"—except he then said such a belief was "biased."[74] One way or another, Lu Xun never said the advent of "True Man" was at hand.

What he said was that his people were subhuman (and surely he really said that we all were, for he saw true men nowhere yet on earth. He meant his works to be a Chinese mirror, but in that mirror all people could see themselves. That is why the works of Lu Xun so clearly belong to world literature). He said that his people were subhuman—in Darwinesque language that was totally un-Darwinian.

But before we can make that clear, we must face, alas, one more level of confusion. If Lu Xun was somewhat inconsistent in his pronouncements on human nature in 1908 and 1918, he was also inconsistent in 1918 alone. He was inconsistent in "A Madman's Diary."

The main argument in "A Madman's Diary" was that it was the beast in human nature that made people eat people. It was the beast in human nature that true men would have to overcome. But when Lu Xun's madman asked himself why even children wanted to eat people, instead of saying it was in their nature, he said, "I know—it is because their mothers and fathers taught them."[75] So the salvation of the human race lay in the salvation of innocent children, children who had not yet learned to eat their kind.

In classical terms, when Lu Xun said man's inhumanity was in his nature, he sounded Xunzian. Xun Zi said man's nature was evil, man's nature was bad for him, but if he could recognize his nature, he could change it. It was not impossible. Even Huxley, as good a Xunzian as Lu Xun, believed that "much may be done to change the nature of man himself," for "the intelligence which has converted the brother of the wolf into the faithful guardian of the flock ought to be able to do something towards curbing the instincts of savagery in civilized men."[76]

But when Lu Xun said man's inhumanity was in his upbringing, he sounded Mencian. Mencius said, "What distinguishes man from beast is very slight; ordinary people give it up, the noble man keeps it." Keeps what? The great man is he who does not lose his *chizi zhi xin*," his "child's heart," his childish innocence.[77] If "the children" could be saved, if they were not taught to eat people, then their salvation lay in their *chizi zhi xin*. They were not beasts by nature. By nature they were fit to be men (yet again, *pace feminae*), "true men."

Lu Xun contradicted himself—and then contradicted himself again. When he spoke of "the nonuniformity of the human race," when he said "there are great differences in the degree to which men have walked the road of evolution. Some have kept the nature of worms and some the nature of apes,"[78] he sounded neither like Mencius nor Xun Zi, but like Wang Chong, who had said in the first century A.D.: "Human nature can actually be good or bad, just as human talents can be high or low." Human nature, like soil, he said, "comes in three grades—superior, medium, and inferior" grades that "cannot be changed."[79] That was why Lu Xun said in 1908 that imperialists were stuck with their "feral nature." That was why he said in 1926, in "On the Advisability of Postponing Fair Play," that counterrevolutionaries were stuck with their "canine nature." "Canine nature," he said, "is not capable of much change. Perhaps after ten thousand years it will be different from now, but what I am talking about now is now."[80]

He almost echoed Edward Livingston Youmans, rabid American Spencerian, who decades before had delivered this famous retort to Henry George, when asked what could be done to fight social injustice—people-eating—in New York:

> Nothing! You and I can do nothing at all. It's all a matter of evolution. We can only wait for evolution. Perhaps in four or five thousand years evolution may have carried men beyond this state of things.[81]

Lu Xun did not ordinarily tell his people, "You and I can do nothing at all." He told them, "You can change. Change from the bottom of your hearts." He told them they could struggle to survive, they could resist imperialists, beat counter-revolutionaries, and change their "national nature." Had he not sometimes believed they could, he himself would have done nothing at all. He wrote, after all, *to change* the national nature.

But he was racked with doubt. All his life he was racked with doubt. He was never sure there was a way out of "the iron room." He was inconsistent. He sometimes sounded sure. But he was not sure. On dark days, and throughout his life there were dark days, he feared it "was all a matter of evolution." He feared imperialists were prisoners of their "feral nature," counter-revolutionaries of their "canine nature," and Chinese of their "national nature"—for "now"—because he shared Youman's view of evolution's speed.

Evolution was too slow. *Evolution* would not "save the children." It would not save our children's children, or their children's children. Children would survive, but their nature would not be changed before our eyes, or their eyes, or their children's eyes. We shall not be changed in the twinkling of an eye ever in evolution.

That was cold Darwinian truth. That *is* cold Darwinian truth, un-shaken even by the efforts of Darwinian revisionists Niles Eldredge and Stephen Jay Gould to speed up speciation. They with their half-justly trumpeted theory of "punctuated equilibrium" have told us, convincingly enough, that "species form rapidly in geological perspective (thousands of years)."[82] But what is geological perspective to us? "It is gradualism that we must reject," Gould tells us, as he points out that "the origin of most species" takes "hundreds of thousands of years," a "geological instant."[83] But is that not gradual enough—for us? If not, Gould tells us that after the "sudden appearance" of species, "in hundreds of thousands of years," they "then persist, largely unchanged, for several million." What if we are stuck in such a period of "stasis?"[84]

Darwin demonstrated the mutability of species. Gould, in his inde-fatigable efforts to convince us that Darwin was wrong in insisting that mutability "must advance by the shortest and slowest steps,"[85] has ironi-cally reminded us that "for all practical purposes," that is, for our purposes, species, especially our own, are, although not immutable, non-mutant.

"Most species," Gould tells us, "exhibit no directional change dur-ing their tenure on earth."[86] New species do, of course, "split off." That is where new species come from.[87] So the "superman" still has a chance. But that "splitting off" can take "hundreds of thousands of years." So Lu Xun was right: The evolutionary advent of the "superman," for us, is surely "too remote." Of course by Gould's reckoning, Youmans and Lu Xun, in their pessimism, were absurdly optimistic. Lu Xun, perhaps with greater Chinese patience, looked for a new nature in ten thousand years, You-mans in four or five. But who could "wait for evolution" even then?

The cold Darwinian truth was that even in Gould's "geological in-stants" of "rapid transition between stable states,"[88] evolution offers us rapid transit nowhere.

Lu Xun vacillated between two views of human nature. In his pessi-mistic moments he spoke the language of what Gould and many others call "biological determinism,"[89] which was also the language of the Chi-nese proverb: *"jiang shan yi gai, ben xing nan yi"* (it is easy to change rivers and mountains, hard to alter one's nature). The protosociobiological bite both of that proverb and of Lu Xun's more pessimistic pronouncements was that by "nature" we were fixed not just in form but in behavior. Lu Xun, naturally, had no notion of the alleged mechanics of sociobiology's "biological basis of all social behavior,"[90] but whenever he spoke of evolu-tion determining any breed's *ben xing* (basic nature or characteristics), he spoke not of physical characteristics, or Darwinian "characters,"[91] of the "trivial" *pimao* sort,[92] but of racial or specific "character" expressed in ac-tion (or, if all was, indeed, determined, in reaction).

If Lu Xun had simply agreed with Huxley, that human nature in general bordered on the beastly, that would have been cause enough for pessimism, but the real bite in his ill-thought-out theory of biological determinism was in his dogma of "the non-uniformity of the human race." Believing that "there are great differences in the degree to which men have walked the road of evolution," believing that "some have kept the nature of worms and some the nature of apes," he did what John Stuart Mill told us not to do. He attributed "diversities of conduct and character to inherent natural differences"[93] within humankind. And so in the evolutionary present, in which all were imprisoned, not only would leopards not lose their spots, but boys would be boys, dogs would be dogs, and people eaters would be people eaters. And so, as they would not change, it was all right to *eat* people eaters, and "beat dogs in the water."

But that was Lu Xun at his most pessimistic. When he was relatively optimistic (very rarely was he absolutely optimistic), he said what his madman said: *"You can change."* He abandoned the language of "biological determinism," for the language of free will—for the age old *Confucian* language of free will.

He abandoned Haeckel. He stood again with Huxley. He stood in this at least, with Nietzsche. But in Chinese tradition, he stood most firmly, although he might never have admitted it, with Confucius. It mattered not that he might sometimes sound Mencian (we can unlearn our unnatural bad habit of eating people) and sometimes Xunzian (we can learn to inhibit our natural desire to eat people), for in either case he was Confucian. "You can change," "you can *ke ji*" (conquer yourselves, or, to please Nietzsche, overcome yourselves).[94] That was *the* Confucian promise.

Confucius did not believe in biological determinism, except perhaps, for a few. Men are "by nature close," he said, "in behavior far apart."[95] "Only the most intelligent and the most stupid cannot change," he said.[96] Most *ren* (men *and* women) *can* change themselves. "I have never seen," said Confucius, "anyone whose power was insufficient."[97]

Xun Zi said, "Anyone who honors Yao, Yu, or the 'noble man,' can change his nature (*hua xing*)."[98] Xun Zi thought man's evil nature needed changing. But he thought *anyone* could change it: "The man in the street can become an Yu."[99] Mencius thought our nature did not need changing. It was good. We had only to obey it. But his promise was the same: "Everyone can become a Yao or Shun."[100] Xun Zi and Mencius agreed: We could make ourselves fit, morally fit—if we would. Will was the way. "It is that you do not, not that you cannot," Mencius told the king who would be King.[101] "Is humanity far away?" asked Confucius. "If I desire humanity, then humanity appears."[102]

The madman's language of free will was Confucian. It was Huxley-ian, Nietzschean, Lamarckian. It was not Haeckelian. It was not Darwinian.

As Asa Gray had seen, the very first time he read *The Origin*, "Darwin makes very little indeed of voluntary efforts as a cause of change."[103] "We shall be changed," in good time, Darwin said, almost as often as Handel,[104] but "you can change" he said hardly ever.

The language of free will did occasionally sneak into Darwin's writing, but it found no safe place there. It lurked in *The Origin*, behind his theory of sexual selection: "Amongst birds . . . successive males display their gorgeous plumage and perform strange antics before the females, which standing by as spectators, at last choose the most attractive partner . . . according to their standard of beauty."[105] It even insinuated itself into his theory of instinct:

A little dose, as Pierre Huber expresses it, of judgment or reason, often comes into play, even in animals very low in the scale of nature. . . . How unconsciously many habitual actions are performed, indeed not rarely in direct opposition to our conscious will! Yet they may be modified by the will or reason.[106]

And in *The Expression of the Emotions in Man and Animals*, Darwin said, "It is . . . possible that even strictly involuntary actions . . . may have been affected by the mysterious power of the will."[107] Nonetheless, the main thrust of that book went quite in the opposite direction.

Darwin wanted to demonstrate "that the chief expressive actions, exhibited by man and by the lower animals, are now innate or inherited,—that is have not been learnt by the individual."[108] The "drooping ears, hanging lips, flexuous body, and wagging tail," Darwin said, of a dog meeting "its beloved master," cannot "be explained by acts of volition . . . any more than the beaming eyes and smiling cheeks of a man when he meets an old friend."[109] For "the far greater number of the movements of expression . . . cannot be said to depend on the will."[110]

Granted, the will even then slipped back into his argument. The will stood godfather to (rebellious) instinct: "Actions which were at first voluntary soon became habitual and at last hereditary, and may then be performed even in opposition to the will." But that was a strange business. Even if we overlook Darwin's double Lamarckian inheritance (he clearly had inherited Lamarck's theory of inheritance), how can we conceive of will—conscious will—evolving before instinct? It is no wonder that Darwin himself confessed: "I have often felt much difficulty about the proper application of the terms, will, consciousness, and intention."[111]

Darwin's vestigial Lamarckianism sometimes blinded him to the most radical implications of his own theory. He himself had said in *The Expression* that "the structure *and habits* of all animals have been gradually evolved,"[112] and in *The Origin* he had said, albeit in the middle of a discussion of "the resemblance between instincts and habits," and the inheritability of "habitual action," that again owed far more than it should have to Lamarck, that "it would be the most serious error to suppose that the greater number of instincts have been acquired by habit in one generation, and then transmitted by inheritance to succeeding generations." For, he said, "I believe that the effects of habit are of quite subordinate importance to the effects of the natural selection of what may be called accidental variations of instincts;—that is of variations produced by the same unknown causes which produce slight deviation of bodily structure."[113] Genetic mutation and natural selection could effect *behavior*.

That is why Konrad Lorenz had a right to say that "the biology of behavior has a special right to claim Charles Darwin as its patron saint"— because Charles Darwin was the first to claim, as Lorenz said, that "behavior patterns are just as conservatively and reliably characters of species as are the forms of bones, teeth, or any other bodily structures."[114] And that, of course, is why sociobiology has a special right to claim Charles Darwin as its patron saint—because sociobiology, as that science's highest priest, Edward O. Wilson has said, has come from an "uncompromising application of evolutionary theory to all aspects of human existence."[115]

"Biology," says Wilson, "is the key to human nature."[116] Biology, Darwinian biology, not philosophy or religion, has at last enabled us "to look inwards, to dissect the machinery of the mind and to retrace its evolutionary history,"[117] which effort has already shown us, says Wilson, not only that "human emotional responses and the more ethical practices based on them have been programmed to a substantial degree by natural selection over thousands of generations,"[118] which is what Darwin himself suggested in *The Expression*, but that there is, indeed, a "biological basis" to "all social behavior."[119]

So Darwin's theory of evolution, says Wilson, has led us to the revolutionary, albeit *(nota bene)* "admittedly unappealing proposition,"[120] that "we are biological and our souls cannot fly free. If humankind evolved by Darwinian natural selection, genetic chance and environmental necessity, not God, made the species,"[121] a species, like all species (to use the catchy but sui-genicidal phrase of another sociobiologist), of "gene machines."[122] An "uncompromising application of evolutionary theory to all aspects of human existence," has led us, says Wilson, to this question—and this conclusion: "If our genes are inherited and our environment is a train of physical events set in motion before we were born, how can there be a truly independent agent within the brain? The agent itself is created by the interaction of the genes and the environment. It would appear that our freedom is only a self-delusion."[123]

Now let us be perfectly clear. Has the patron saint of sociobiology *proven* its high priest right? Has Darwin *proven* human "freedom" a "self-delusion?" Not at all. No one has ever come close to proving that free will does not exist, any more than anyone has ever come close to proving that it does. But Darwin's theory of evolution logically leads to determinism—and nowhere else—not to a simple "biological determinism," but to a determinism born of the chance "interaction" of "genetic chance and environmental necessity." Darwin explains no other forces behind the evolution of anything. He *mentions* will, "the mysterious power of the will," but *nothing* in his theory tells us how such a power could be.

Darwin admitted as much. He said in *The Origin*, "I must premise, that I have nothing to do with the origin of the primary mental powers, any more than I have with that of life itself."[124] The origin of matter, *wu zhong sheng you* (the birth of being from nonbeing), the energizing of energy, the leaping of *ta stocheia* into life, the building up of senseless blocks into things with senses, if not sense, these are not just things that we do not yet understand, they are things almost literally inconceivable. They can probably only be conceived if we simply give up and accept "energy," and then say that there is no divide between the element and the compound, the inorganic and the organic, the unconscious and the conscious. All is just energy in motion determined by the motion of energy. But that does not explain free will. Free will is either illusory or metaphysical. Therefore, it either is not, or it is not Darwinian. Any honest, downeaster Darwinian, if asked the way to free will, would have to reply, "You can't get there from here."

Now lest theists leap in at this point with too smug an "a ha!," we should admit that theists cannot explain free will either. They can perhaps more readily accept it, as a "given," but they cannot explain how it could be given, for we cannot conceive how *creatures*, natural or divine, could be free from their creator, how naturally or divinely wound-up toys could have a will of their own, how puppets could move with no strings attached. Pinocchio's life, our life, if we have life, defines all logic. So what do we do? We accept determinism and deny our existence, or we accept our existence existentially: *volo ergo sum*.[125] But Darwin was not an existentialist, theistic or atheistic. He accepted life, but everything he explained about it, descent with modification, he explained in the language of determinism, not in the language of free will. So the language of free will is *not* Darwinian.

Stephen Jay Gould, one must admit, has recently tried to rescue Darwin from the camp of the biological determinists—who in their most potent current reincarnation are, of course, the sociobiologists (Haeckel's heirs, though they might not admit it.) Gould has met, however, with uncertain success.

A champion Darwinian champion, for all his notoriety as a Darwinian revisionist, Gould hates biological determinism. That is why the last chapter of Wilson's *Sociobiology* leaves Gould "very unhappy indeed."[126] For Gould refuses to believe that "the statement that humans are animals" implies "that our specific patterns of behavior and social arrangements are in any way directly determined by our genes."[127] There is, he says, no "direct evidence" whatsoever "for genetic control of specific human social behavior."[128] There are no "specific genes for specific behavioral traits." So "against the idea of biological determinism," Gould champions "the concept of biological potentiality."[129]

But what is that? *At first glance,* "the concept of biological potentiality" looks like a new imperial cloak for the doctrine of free will. All is biological. All is perfectly natural. But our biology has given us "flexibility." Evolution, by increasing our brain size, has "added enough neural connections to convert an inflexible and rather rigidly programmed device into a labile organ." Evolution has given us "a brain capable of the full range of human behavior and rigidly predisposed toward none."[130]

Granted, says Gould, "the range of our potential behavior is circumscribed by our biology,"[131] but within that circumference we live "as flexible animals with a vast range of potential behavior."[132] For "the brain's enormous flexibility permits us to be aggressive or peaceful, dominant or submissive, spiteful or generous."[133] Gould almost echoes de Tocqueville: "Providence has not created mankind entirely independent or entirely free. It is true that around every man a fatal circle is traced beyond which he cannot pass; but within the wide verge of that circle he is powerful and free."[134]

Gould, of course, wants no truck with "Providence," not the Providence of old at any rate. Nature is the great provider. So we are still perfectly natural, however we behave. Gould echoes Huxley's dictum: "The thief and the murderer follow nature as much as the philanthropist."[135] "Basic human kindness," says Gould, "may be as 'animal' as human nastiness."[136] "Violence, sexism, and general nastiness," he says, "*are* biological since they represent one subset of a possible range of behaviors. But peacefulness, equality, and kindness are just as biological."[137]

Gould wants his potentiality and his biology too: "Our genetic make up permits a wide range of behaviors [sic]—from Ebenezer Scrooge before to Ebenezer Scrooge after."[138] Our genetic makeup *permits* a wide range of behavior. It does not *determine* our behavior. Gould seems to offer us a choice.

But does he really? If we take a second glance—through "biological potentiality's" new, imperial, free-will shirt—we will still behold determinism. *Why* do we behave like Scrooge after or Scrooge before? What *causes* us to behave one way or the other?

> Upbringing, culture, class, status, and all the intangibles that we call 'free will,' *determine* how we restrict our behaviors from the wide spectrum?—extreme altruism to extreme selfishness—that our genes permit.[139]

> The influences of class and culture far outweigh the weaker predispositions of our genetic constitution.[140]

If upbringing, culture, class, and status determine what *we* do, then we do nothing. Nature, nurture—what difference does it make? Genes? Society? Who cares who pulls our strings? Cultural determinism is just as destructive to us as biological determinism. We cannot *be* without free will. If there is *no* circle within which we are "powerful and free," then truly: "Existence is but an illusion."[141]

Does Gould save us by including in his formula "all the intangibles that we call 'free will?'" No—for look at what "free will" means to him: "Chanciness . . . gives meaning to the old concept of human free will." Does it? Not at all. "Chanciness" reflects nothing but "the genuine randomness of our world." Our "behavior" is unpredictable because we are pushed and pulled at every turn by "a concatenation of staggering improbabilities."[142] "Chanciness" does *not* set us free.

Wittingly or unwittingly, Gould has backed from Wilson's camp into Skinner's, or in classical Chinese terms, into Gao Zi's. In rejecting Wang Chong's contention that we are good, bad, or indifferent, at first, by nature, he has accepted Gao Zi's counter claim that by nature we are malleable. We can be shaped. That is our "biological potential":

> Gao Zi said: "Nature is like willow wood. Goodness is like a wooden cup. Making goodness from human nature is like making a cup from willow wood."[143]

A second time he switched metaphors, but the result was the same:

> Gao Zi said, "Nature is like flowing water. If you dig a ditch to the east it will flow to the east. If you dig a ditch to the west, it will flow to the west. Human nature is no more disposed towards good or evil than water is towards east or west."[144]

It is all in the channeling. Our "biological potentiality" "permits a wide range of behaviors" because by nature we are susceptible to a wide range (in Skinner's phrase) of "behavioral engineering."[145] That is why Gould believes that "our biological nature does not stand in the way of social reform."[146]

Gao Zi was a proto-Skinnerian. Gould took his stand with Gao Zi, and with Watson, and with Skinner himself momentarily, only to fall over backward with the rest of them into the same heffalump trap.

At *third* glance (we are now down to the emperor's new undershirt), we see free will again. Skinner did his best to deny free will. "You can't have a science," said his alter ego in *Walden II*, "about a subject matter which hops capriciously about."[147] But he failed to explain the *choices* of his alter ego. He failed to explain himself. For he *chose* the direction of his "behavioral engineering," just as Gould would choose the direction of his "social reform." Once again, Gould says:

> Violence, sexism, and general nastiness are biological since they represent one subset of a possible range of behaviors. But peaceful-ness, equality, and kindness are just as biological—and we may see their influence increase *if we can create* social structures that permit them to flourish.[148]

Thus "biological potentiality" is revealed as a pseudo-scientific term. Above it, through that word "create," Gould himself lets free will sneak back in, still unexplained, still out of place in Gould's own Darwinism, but there. For Gould makes a choice, and he asks us to make a choice, and what a choice, the choice between "Ebenezer Scrooge before" and "Eben-ezer Scrooge after," the choice between good and evil, the oldest choice in the Book—and the choice that Lu Xun would have us make:

Nimen keyi gai le—You can change.

To eat people, or not to eat people? Absolutely inexplicably, Lu Xun said it was up to us.

Lu Xun wavered back and forth between two theories of human na-ture, two contradictory theories, each also contradictory within itself, one deterministic, one voluntaristic (although with a deterministic side), one pessimistic, one optimistic (although with a pessimistic side), both sup-ported *with* Darwinism, neither supported *by* Darwinism.

Lu Xun's pessimistic, deterministic theory of human nature made men products of nature, different products with different natures, natures determined by evolution, natures fixed by the forces of evolution. So men could not change their spots. Men had evolved from beasts, so men had beastly natures or semibeastly natures. People-eaters, imperialists *and* all too many Chinese, had beastly natures, and they would keep their beastly natures until natural selection took them off the stage of natural history.

This theory was "pessimistic" in three ways. First it left no room for *ren wei* (human action): A people's fate was "in its genes." But *that* idea was "pessimistic" because Lu Xun seemed to say, when he was in his pessimistic mood, either that Chinese were genetically fit only to be food for imperialists, or that they were genetically fixed to be people-eaters themselves, of the worst kind—eaters not of others but of their own, doomed to eat each other up until "true people" ate *them* up.

The biological determinism of that theory was Darwinesque, but it was not Darwinian, because it was teleological and idealistic. Evolution would create "true men," if necessary over the dead bodies of the Chinese, but "true men" were not defined biologically. They were not described as a *fitter* species, but as a *better* species.

Lu Xun's optimistic theory, of course, was even less Darwinian—because it was voluntaristic. One's fate was not in one's genes, it was in "one's own hands." People could be "true people" if they wanted to be. That was no Darwinian promise. If anything evolutionary, it was Lamarckian. But actually it was *not* Lamarckian, because again it was not biological. When Lu Xun spoke of "true men," he was no more speaking of a new biological species than Nietzsche was when he spoke of supermen. Granted each sounded as if he were speaking of a new species, but true men and supermen were to distinguish themselves from men not in form but in behavior, and that behavior was not dependent on any change of genetic structure, because we could *already* be "true men"—if we wanted to be. Lu Xun contradicted himself: He did say "true men" would be different from "men," in nature. But he also said it was within our nature to be true men, it was within our nature to change our nature. So what sort of nature was that?

Lu Xun's optimistic, voluntaristic, idealistic theory of human nature was *not* Darwinian. What was it? Once again—it was Confucian. For what was a "true man?" A *zhen de ren* was a *ren* who would not *chi ren*—a person who would not eat people. A *zhen de ren* was a *ren ren*, a humane man: "*ren zhe ren ye*"—to be human, truly human, was to be humane.[149]

8

Evolution and Ethics Again

So Lu Xun's pronouncements on human nature drag us back to evolution and ethics. Lu Xun said that "eating people" was natural. He also said, with Flanders' and Swann's "reluctant cannibal," that "eating people is wrong."[1]

But how could it be both?

When Lu Xun said that "beastly natured patriots," imperialists, *could* not change their spots, he refused to accept the Darwinian logic that Liang Qichao accepted (for awhile):

> If a country can strengthen itself and make itself one of the fittest, then, even if it annihilates the unfit and the weak, it can still not be said to be immoral. Why? Because it is a law of evolution. Even if we do not extinguish a country that is weak and unfit, it will be unable to survive in the end anyway. That is why violent aggression, which used to be viewed as an act of barbarism, is now viewed as a normal rule of civilization.[2]

> When European countries meet with other European countries they all take reason to be force, but when European countries meet non-European countries, they all use force for reason. And so they must, because of evolution. The struggle for existence makes it natural. So what cause can there be for blame? What cause can there be for hate?[3]

Many found that logic hard to take, but rejecting it, rejected Darwin:

> Oh, how I dread
> To talk of evolution!
> For if fit flourish and unfit fail,
> Then whom do we dare hate?[4]

Lu Xun, however, dared to hate, but did *not* dread to talk of evolution. He dared to say what Huxley said, that "ape" and "tiger" are *naturally* within us, but we must struggle to repress our "ape and tiger promptings,"[5] that Evolution orders us to eat people, but we must not take evolution's orders—because those orders are our evolutionary endowment of "original sin,"—and we should not sin.[6]

But how in evolution's name can "sin" exist?

Sin cannot exist unless we do, and good does. But evolution cannot explain the existence, the *real* existence, of either. If all we do we do naturally, we do nothing—and are not. If natural law rules all, there is no good or evil. Lu Xun's naturalism, his materialism, was totally at odds with his moral indignation.

Lu Xun, in his early thought, ran into philosophy's two greatest questions, Do we exist? and Does good exist? But he never recognized them. He never asked them. He answered them without asking them—because he was not interested in "philosophy" (in that too he was like Confucius). He did not see the philosophical contradictions in his own thought any more than he saw the philosophical contradictions in the thought of his two Darwinian mentors, Huxley and Haeckel, or the philosophical contradictions *between* the thought of his mentors, Huxley and Haeckel. He did not appreciate the irony of Haeckel's racial pride, held firmly in the face of his monist denial of free will. He did not appreciate the irony of Huxley's leap of faith, over the seemingly healthy body of his agnosticism. Huxley himself never fully appreciated that irony, though he was philosopher enough to glimpse it. He admitted that it seemed illogical to maintain that man, "as much a part of nature, as purely a product of the cosmic process as the humblest weed," should yet be able to rebel "against the moral indifference of nature" and combat the cosmic process.[7] But damning logic, he maintained that just the same. And that is what Lu Xun maintained, though never damning logic.

Lu Xun, in his relatively optimistic moments, accepted our existence as readily as an existentialist. He never consciously entertained the strange Sartrian notion that "man is condemned to be free,"[8] but he granted us freedom. Without ever facing the implications of that act of choice, he granted us choice of action.

He also accepted, without admitting it, the existence of good—because he demanded of us moral action.

We can see that demand in his double opposition to imperialism—in his call to his countrymen not just to resist imperialism, but to reject it.

If Lu Xun had merely urged his countrymen to resist the White Peril, as so many other patriotic writers did, reminding them that only the fit survive, we might still say that he urged his countrymen to heed evolution's one imperative: struggle to survive (though even that "imperative" could not be truly evolutionary—for if all is matter only, nothing can order

and nothing obey). But Lu Xun did not just urge Chinese to resist imperialists. He urged them in their self-defense (which he was for—he was no
pacifist[9]) to reject the *idea* of imperialism, to reject it even as a means of
self-defense. He denounced Chinese who took pride in Western talk of
the Yellow Peril and dreamt of walking victorious through the streets of
London.[10] A decade before Lu Xun's madman cried, "You must realize
that the future will have no place for people who eat people,"[11] Lu Xun
told his people to "defend themselves" but not to "savagely eat the weak
of the world,"[12] to "preserve themselves" but not to "seize and swallow
up guiltless countries."[13]

Why not? Why not eat people in a world of wild men "red in tooth
and claw?" Lu Xun asked the question and answered it:

> Our land has long suffered violence. China is not yet a corpse, but
> vultures already gather round. We have not only lost territory, we
> have lost great wealth, and people have suffered and died. Nonethe-
> less if we must now have sharp weapons and stout shields, it is to
> defend ourselves on all sides, not to become "wild pigs and pythons
> and devour other countries." We act only in self defense. We do not
> imitate aggressors. We will not commit aggression against others.
> We do not approve of aggression. —Why not? Because we know
> what it means to us. We are the enemies of animal nature.[14]

Sixty years before Zhou Enlai reassured the world that China (unlike
certain other countries) would *"bu cheng ba"* (not become a hegemon),[15]
Lu Xun renounced aggression. His "good grounds" for doing so were not,
however, proto-Marxian (Marxists, indeed, have no good grounds for renouncing aggression). They were quintessentially Confucian. It was the
Confucian Golden Rule that he cited as his good grounds for denouncing
and renouncing beastly, cannibalistic imperialism. The phrase he used
above, *"fan zhu ji,"* (to look to one's own experience for knowledge of
something) comes from *The Analects*, where it led to Confucius' negative
statement of the Golden Rule, "Do not do unto others what you would
not have them do unto you," Confucius' rule of *shu*, the virtue, so he said,
that tied all his teaching together.[16]

But the Golden Rule does *not* give good grounds for not eating people. Knowledge that one does not want others to eat one does not explain
why one should not want to eat others—much less why one should not
eat others if one wants to eat others. Indeed, knowledge that we do not
want others to eat us might well lead us to eat others *lest* others eat us.

So Lu Xun's first answer to his question, why should we not commit
aggression, was no answer at all.

He might, of course, have made it one. He might, indeed, have made it an "evolutionary answer" to his question, if only he had made of his golden rule a ploy, or a "strategy,"—to use the highly unscientific language of the sociobiologists.

Twenty-four hundred years ago, Mo Tzu made a strategy of the golden rule. He said: "I must first set myself to loving and profiting the parents of others, and then they will repay me by loving and profiting my parents."[17] Such filially pious "altruism" might be harder for a sociobiologist to explain than "altruism" aimed at one's offspring, or one's peers, or one's peer's offspring, but the "principle" in any case would be the same: the golden rule would be employed as a ploy, as a "pact,"[18] as a form of "barter":[19] "You scratch my back and I'll scratch yours,"[20] or "I'll help your genes *so that* you'll help mine." That is what Trivers, Wilson, and Dawkins all call "reciprocal altruism,"[21] the reflection, in the illusory "superstructure" of our "consciousness," of a naturally selected survival mechanism in the unconscious substructure of our genetic reality.

Mo Zi did not need any biologists or sociobiologists to teach him the principle of reciprocal altruism. Lu Xun did not need any biologists or sociobiologists to give him a protosociobiological answer to his question, why should we not eat people. He might, with Mozian ease, have said, in effect, what Dawkins has said: "A cannibal strategy would be unstable for . . . there is too much danger of retaliation."[22] —But he *did not.*

Why should we not eat people? —Because, he said, "We are the enemies of animal nature."[23] —Another nonanswer. "If we want to surpass the birds and beasts," he said, "We must not envy the ambitions" of "beastly-natured patriots."[24] But why not? *Why* should we be "enemies of animal nature?" *Why* should we "want to surpass the birds and beasts?" —Because "eating people is wrong." Those were not the madman's own words, but that is what the madman said, and that is what Lu Xun said, in *almost* everything he ever wrote.

But how did he *know?* How did he *know* that eating people is wrong? He knew, if he did know, for we cannot know for sure that he knew, through *intuition*, Huxley's word, that oddest of words for an agnostic.

Huxley said in *Evolution and Ethics*:

The propounders of what are called the "ethics of evolution," when the "evolution of ethics" would usually better express the object of their speculations, adduce a number of more or less interesting facts and more or less sound arguments in favor of the origin of the moral sentiments, in the same way as other natural phenomena, by a process of evolution. I have little doubt, for my own part, that they are on the right track; but as the immoral sentiments have no less been evolved, there is, so far, as much natural sanction for the one as the other. The thief and the murderer follow nature just as much as the philanthropist. Cosmic evolution may teach us how the good and the evil tendencies of man may have come about; but, in itself, it is incompetent to furnish any better reason why what we call good is preferable to what we call evil than we had before. Some day, I doubt not, we shall arrive at an understanding of the evolution of the aesthetic faculty; but all the understanding in the world will neither increase nor diminish the force of the *intuition* that this is beautiful and that is ugly.[25]

Huxley and Lu Xun were both *"guilty"* of what Wilson scornfully calls, with the acquired characteristic grace of a social scientist, *"ethical intuitionism."*[26] Huxley in the passage above, almost admitted it, although, alas, he failed to face that admission philosophically. Lu Xun did not admit it, although he applauded Huxley's faith in intuition in another (surprising) context ("England's Huxley maintains that [scientific] discovery, gets its start in intuition".[27] Nonetheless, Lu Xun *was* "guilty" (although *nota bene,* Wilson: Guilt too sets us free). On almost every page he ever wrote, Lu Xun revealed a *zhengyigan,* a "sense of justice *(dikaiosyne)"* that would have done justice to an ancient Greek.

Lu Xun's *zhengyigan* was the font of all his work. Like Huxley, he thought it self-evident that "whether we look within us or without us, evil stares us in the face on all sides."[28] In every short story, in every essay, in every serious poem and letter, beneath all his great, subtle and insinuative craft, Lu Xun pointed out evil and condemned it. How did he recognize it? He simply knew it when he saw it—or he thought he did, because he had a *zhengyigan,* a sense of justice.

But what is a *zhengyigan?* We should not fool ourselves, or others: We do not know. It is either illusion or a glimpse, by an incomprehensibly liberated bit of matter, of a transcendent truth. So Lu Xun's *zhengyigan* brings us back to the great materialist-idealist divide. And we must pay our money and take our choice, or throw up our hands in despair.

Materialists of many stripes have indeed, mounted a formidable modern attack on "ethical intuitionism." Haeckelians, Freudians, behaviorists, sociobiologists, bedfellows all, have all done their best to prove that ethical intuition is illusion, though none have given up making ethical judgments of their own. Marxists, too, the most militant of materialists, ought to say that ethical intuition is illusion, and yet they have never stopped talking of justice. The lifeblood of Marxist thought is Marx's *zhengyigan.* The lifeblood of Mao Zedong's thought was his *zhengyigan:* "Marxism-Leninism can be boiled down to this truth: To rebel is justified."[29] "Wars in history are of two kinds: one is just, one is unjust."[30] Even "proletarian consciousness" is ultimately consciousness of justice. So Chinese Marxist historians, or Chinese historians writing under Marxist constraints, have seen nothing odd in their praise of Lu Xun as one who "steadfastly held to a just *(zhengyide)* position"—maintaining that there could and should be no compromise between right and wrong, good and evil, and revolution and counterrevolution.[31]

But do materialists who so sense justice not become idealists? A sense of justice is nonsense unless justice exists to sense. Ethical intuition is nonsense unless there is something to intuit. Sartre said that "every man who sets up a determinism is a dishonest man."[32] Perhaps every materialist who sets up a sense of justice is also a dishonest man, or woman. But there is no settling the matter. It is dreadfully hard to believe in the existence of good, and dreadfully hard not to. Nonetheless, one thing is clear—and important. Regardless of whether or not one can (to one's own satisfaction) explain it or explain it away, Lu Xun did have a *zhengyigan*. He thought that he had an intuitive sense of good, and that, whatever it was, is what drove him to write. Therefore, although materialists, Haeckelian, Freudian, sociobiological, or whatever, may still endeavor to explain Lu Xun's thought materialistically, the label "materialist" for Lu Xun himself simply will not stick.

Lu Xun was idealistic enough, of course, in his championship of his "Mara," or Satanic, poets, Byron, Shelley, Keats, Pushkin, Mickiewicz, and so on, "spiritual warriors," men of "genius," who struggled against the crowd to change the world.[33] He was idealistic enough in his exhortation to "respect individuality and foster spirit,"[34] and in his protestation that "scientific discovery is often propelled by a supra-scientific power, or, to put it differently, by a non-scientific ideal."[35] But never was he more idealistic than in his expression of his sense of justice.

Two questions remain, however, about that sense of justice. What sort of a sense of justice was it, and (and this is the most difficult question—but if Lu Xun's works are worth reading, how can we not ask it) how *just* was his sense of justice?

The first is difficult enough—and already dangerous. When all is said and done, the best label for Lu Xun's sense of justice is "Confucian," not Darwinian, not Huxleyian, not Haeckelian, not Buddhist, not Daoist, not Marxist. But if we say that "Confucian" explains Lu Xun's sense of justice, or even if we say that a combination of Chinese and Western influences explains it, we walk into a cultural determinism as devastating to Lu Xun's thought as any biological determinism. We do not have to get down to Wilson's or Dawkins' "selfish genes." If Lu Xun thought what he thought simply because he was programmed by his culture, his class, his upbringing, or his education to think what he thought, then he did *not* think what he thought. He was thoughtless, and so are we, and so our effort to understand his thought is a double delusion.

Yet, "verily," as Asa Gray said about something else, "if this style of reasoning is to prevail—'thinking is but an idle waste of thought, and naught is everything, and everything is naught.'"[36] So I will not use "Confucian" as a causative label, but only as a descriptive one: to insist that Lu Xun was Confucian in his moral stance—even in his moral stance against Confucian morality.

His great indictment of traditional Confucian morality, in "A Madman's Diary", was a Confucian indictment. He said that throughout all the annals of Chinese history, two stark words, "eat people," stared out from behind the Confucian words *ren yi daode* (humanity, justice, and morality), but in saying that, he did not *attack ren yi daode*, he attacked *"ren yi daode,"* the false, hypocritical "morality" that masked cruelty and greed, and the innocent but ignorant "morality" that masked murderous, or suicidal, superstition. He attacked most vehemently *"zhengren junzi"* (righteous gentlemen) because they were not righteous gentlemen, just as Confucius attacked *xiang yuan* (village "good people") because they were *de zhi zei* (thieves of virtue).[37] Lu Xun attacked *"renrenmen"* (humane people) because they were not humane people,[38] because he longed, as much as Confucius, for a world in which people would be humane, because, as we have seen, he believed that only "humane people" were "true people."

Lu Xun was like Ji Kang, one of his few Chinese heroes, the third-century leader of "The Seven Sages of the Bamboo Grove," Ji Kang who, by his own confession, was "always criticizing King Tang and King Wu and denigrating the Duke of Zhou and Confucius," but whose works seethed nonetheless with Confucian moral indignation, against the hypocritical, orthodox supporters of the "religion of the day," who were supporters also of the Sima clan, which was slowly usurping power in Ji Kang's state of Wei.[39] Ji Kang, in the end, figuratively left the Bamboo Grove, to speak out for a falsely condemned friend—in that act alone showing himself a better Confucian than the sycophantic "upholders of the faith," who then condemned Ji Kang himself, because he would not join them.

Ji Kang knew himself. He knew all along he was in danger, because, as he said, "I uncompromisingly hate evil and recklessly speak straight."[40] Lu Xun, master of satire and insinuation, may not always have seemed to speak "recklessly" or "straight." Enemies taunted him, saying, "He only beats lap dogs, he does not curse warlords." But Lu Xun said it was dangerous enough cursing lap dogs, for "Who is it who pays attention when you curse out warlords? Warlords do not read magazines. They rely on their lap dogs to sniff about, and on their backup lap dogs to bark."[41] And "lap dogs are often even meaner than their masters."[42] In any case, he said, he was not going to be suckered into suicide.[43] He was going to engage in "trench warfare."[44]—But in the trenches, against warlords and the Guomindang, he was still in danger, and so spoke out quite recklessly and straight enough, because he too "hated evil."

Consider only one example of his recklessness, the pair of lectures he delivered in Guangzhou (Canton) on July 23 and 26, 1927, on the archly "academic" topic of "The Relationship between Drugs and Wine and the Writing and Spirit of the Wei-Jin Period"—Ji Kang's period.[45]

These lectures require more than a word of explanation. Lu Xun was in Guangzhou in part because he had spoken out too recklessly in Beijing, after what he called "the darkest day since the founding of the Republic,"[46] March 18, 1926, the day the reigning warlord in Beijing, Duan Qirui, ordered troops to fire on student demonstrators (whose demonstration had started at Tiananmen), killing forty-seven and wounding over one hundred and fifty.[47]

In a series of angry articles (the most famous written in memory of his student, Liu Hezhen) Lu Xun did *not* content himself with beating the academic "lap dogs" who supported Duan Qirui. He openly cursed the warlord himself.[48] Fearing, however, that he might indeed have been too reckless, he went into hiding, taking refuge, for a month, in a series of hospitals.

In May, he went home, safely enough, as things turned out. In August, nonetheless, he left Beijing, "scandalously" taking the same train south as his favorite student, Xu Guangping ("destined" to become, a year later, sans divorce or marriage, his second wife). Traveling together only to Shanghai, they there parted, temporarily, she going to Guangzhou and he to Xiamen—to Xiamen University. He taught there for a term, enduring "collegial" squabbles, and that famed absence that makes the heart grow fonder. Then he too went to Guangzhou, to teach at Zhongshan University, with Xu Guangping as his assistant.

But Guangzhou proved more dangerous than Beijing.

On April 12, 1927, in Shanghai, Jiang Jieshi (Chiang Kaishek), military leader of the Guomindang's Northern Expedition against the warlords, turned against his alleged allies, the Chinese Communists, and did his best, with the help of the Shanghai Green Gang, to slaughter them to the last man and woman.

On April 15, Qian Dajun, on Jiang Jieshi's orders, made a similar attempt at Guangzhou, slaughtering, it has been said, some three thousand "Communists, workers, and progressive students."[49] Lu Xun, allerted that morning that he was attacked in posters at the university, was urged by friends to flee. But instead he went to the campus to an emergency meeting of department chairmen and tried to convince the university to defend its students. Failing, he resigned.[50]

His fame saved him from being arrested himself. It also led the university, vainly, to attempt to woo him back. And in July, it led the city government, not without malice aforethought, to invite him to take part in a summer lecture series—to put him on the spot. Ouyang Shan, the student (already kicked out of the university) who took down what Lu Xun said, has said in a memoir, not unreasonably, that the government's invitation was a setup, either to win praise for itself for its "liberalness," or to lure Lu Xun into statements seditious enough to warrant his arrest.[51] Lu Xun recklessly responded with his literary lecture on the Wei-Jin Period.

Recklessly indeed, he condemned the Nationalists' "party purge," through very thinly veiled criticism. Ouyang Shan, in 1979, most significantly, praised Lu Xun for *"jie gu feng jin,"* for "satirizing the present with the past," the very "crime" for which Lu Xun's spiritual desciple, Wu Han, was killed in the Cultural Revolution.[52]

Lu Xun was surely "guilty" of "satirizing the present with the past." Through stories of the third-century despots, Cao Cao and Sima Yi, who killed Kong Rong and Ji Kang, he satirized Jiang Jieshi. Through stories of third-century "Confucians" who in Confucius's name trampled on the principles of Confucius, he satirized Nationalists who in Sun Yatsen's name trampled on the principles of Sun Yatsen. Conversely, he called those accused of opposing Confucius, or Sun Yatsen, their "true followers." Ji Kang and his friends, he said, were like "true followers of Sun Yatsen," who "scowl when they hear people talk hypocritically" about Sun Yatsen's Three People's Principles.[53]

In satirizing the present, however, Lu Xun made a most perceptive observation about the past:

> Ji Kang and Ruan Ji have always been accused of defiling the Confucian faith. But in my opinion, that verdict is erroneous. In the Wei-Jin Period, those who professed the Confucian faith looked pretty good, but actually defiled it. Those who superficially defiled it, actually confessed it. They truly believed it. . . . The so called opponents of the Confucian faith . . . were actually old-fashioned gentlemen who treasured the Confucian faith.[54]

Lu Xun saw through Ji Kang's and Ruan Ji's iconoclasm. Did he not see through his own?

The moral indignation of Ji Kang and Ruan Ji, leaders of the Seven Sages of the Bamboo Grove, was Confucian, and so was the moral indignation of Lu Xun.

But although his moral indignation was Confucian, it still led to a contradiction, to the great contradiction, or, at very least, to the great seeming contradiction that every reader of Lu Xun's works must face—before asking the ultimate question: How moral was Lu Xun's moral indignation, how just his sense of justice, how *zheng* his *zhengyigan?*

That is the ultimate question, that we each must ask, armed though we are only with our own *zhengyigan,* tempered, at best, with our own powers, such as they be, of reason, because moral judgment is the essence of Lu Xun's work, moral judgment not just for his day but for ours.

We should, of course, judge his judgment in his every essay, story, poem, and letter. That is why we should read Lu Xun and not just books about Lu Xun.[55] His works remain worth reading because his moral judgments speak to questions much too much still with us. But over all those individual moral judgments two general moral judgments sound, and two injunctions, that all readers must confront.

And so we come, at last, to "the great contradiction."

Lu Xun told his people two things: Stop "eating people," and start "beating dogs who have fallen into the water."

"Eating" meant anything from "hurting" to "killing." "Beating" meant anything from "cursing" to "killing." "Dogs" were people who ate people. So Lu Xun told his people: Eating people is wrong, but eating people who eat people is right.

That was not an astonishing position. Nor was it a position Lu Xun reached only after a slow intellectual evolution. It was definitely not a position he reached only after Marxist-Leninist enlightenment. He pronounced it most clearly in 1925, in "On the Advisability of Postponing Fair Play." He pronounced it quite clearly enough in 1918, in "A Madman's Diary," when he said that "true people," people who did not eat people, would one day rid the world of people who did eat people, the way they would rid the world of wolves and insects. He justified it, with Darwinian arguments on both sides, as early as 1908.

Darwin told Lu Xun, as we have seen, that people-eaters were evolutionary low life bound for extinction, that true people were evolving who would not eat people. Darwin also told Lu Xun that evolution was no dinner party.[56] So Lu Xun concluded that people opposed to people eating could not be soft on people-eaters. They could not in the struggle for existence, even though true people were bound to triumph in the end, be too humane in their struggle with the inhumane.

Was there "a great contradiction" in that position? Officials in the People's Republic would say, of course, "Of course not." And "realists" round the world would say, "Amen."

It is obvious why the answer is obvious in the People's Republic. Justification of the revolution depends on it. And that justification is not just cynical self-justification. American armchair pacifists—even tested pacifists, who have tasted tyranny, cannot lightly scorn Mao Zedong's most honest conviction that "to rebel is justified." Let them compare the Manchus and the Warlords and Jiang Jieshi with George III. Americans had much less cause to take up arms than the Chinese. We cannot *lightly* scorn Mao Zedong's defense of "inhumanity":

> "You are inhuman." Quite so. We refuse to govern humanely when we deal with the reactionary behavior of the reactionaries and the reactionary classes.[57]

> We cannot love our enemies. We cannot love the ugly things in society. Our goal is to wipe out such things. That is human common sense.[58]

Quite so. And it has been so throughout history.

Is that what Lu Xun said? Yes. He was no pacifist, and it would be silly for us to express shock that he was not. Slow though he himself was to revolutionary action, slow, indeed, though he was to urge others to revolutionary action,[59] he nonetheless rejected pacifism early and late. "Struggle?" he said in 1928, "I think it is right. If people are oppressed, why not struggle?"[60] Tolstoy's pacifism, he said in 1908, was a noble ideal, but impractical. To defend themselves against "wolves and tigers," Chinese must take up arms.[61]

"Wolves and tigers," of course, were imperialists. But the "dogs" that Lu Xun thought Chinese should beat, in 1925, were Chinese.

Moreover, in advocating "the postponement of fair play," in beating those dogs, Lu Xun went well beyond the rejection of pacifism.

In 1921, in "Ah Q zheng zhuan" (The true story of Ah Q), Lu Xun had already made clear his belief that the Revolution of 1911 had failed because it had not wrested power from the ruling, people eating, gentry class. When the revolution was over, the Manchus were gone, but the Chinese who had held local power under the Manchus held power still—and still ate people. "When the revolution finally began," said Lu Xun in 1925, "the whole pack of gentry with their stinking pretensions immediately became as scared as stray dogs, and coiled their queues on top of their heads. The revolutionaries with their modern ways—the modern ways that the gentry had formerly detested—behaved in a 'civilized' manner indeed, saying, 'All shall be reformed.' We do not beat dogs in the water. Let them climb out. So they climbed out, lay low until the latter half of 1913, the time of the Second Revolution, and then burst out to help Yuan Shikai bite a host of revolutionaries to death, and China once again sank day by day into darkness. And so it is to this day. Do not talk of 'old loyalists.' Look how many 'young loyalists' there are. This is because the good will of our revolutionary martyrs, the mercy they showed to demons, has let those demons multiply, so that hereafter clear headed youth, to resist the darkness, will have to expend even more, even more, energy and lives."[62]

That is why Lu Xun came to conclude that Chinese should "postpone 'fair play.'" If decent people mistake 'loosing the evil' for leniency, and insist on being tolerant," he said, ". . . then this present state of chaos could last forever."[63]

Was that some revolutionary truth that one could only grasp with Marx's help? Did Lu Xun's grasping of it mark a milestone on the path of his gradual enlightenment? Again, of course not. It was a cry as old as the hills, at opposite ends of the earth. Mo Zi, for one, had said what Lu Xun said, twenty-four hundred years ago. He could not love his enemies either, even though he was an apostle of "inclusive love *(jian ai)*" and an enemy of the Confucianists' alleged exclusive love. Somewhat oddly, given those circumstances, he attacked the Confucianists, as he saw them, for being soft on violence:

> If a victorious general uses Confucian tactics and orders his troops not to pursue a fleeing enemy, not to shoot the enemy when they are in straits, but to help them with their chariots if they get stuck, then violent people will get away alive, and he will fail to rid the world of the harmful.[64]

Mo Zi was as "god-fearing" a man as Moses, but neither was for misguided mercy. Moses, in victory, chastised his officers for sparing enemy women and children—in language that makes him sound like a "selfish gene machine":[65]

Have you saved all the women alive?

Behold, these caused the children of Israel, through the counsel of Balaam, to commit trespass against the Lord in the matter of Peor, and there was a plague among the congregation of the Lord.

Now therefore kill every male among the little ones, and kill every woman that hath known man by lying with him.

But all the women children, that have not known a man by lying with him, keep alive for yourselves.[66]

For all his Haeckelian Darwinism Lu Xun never went that far—calling for the beating of dogs in the water, he did not call for the drowning of puppies (although he did register worry about the proliferation of "young loyalists"). Were we wrong, however, to call Lu Xun a disgruntled Confucian? Was he really a Moist?

On this issue, the distinction gets so blurred, with Confucians and Moists both at their own throats, that it is hard to say. Some say Mo Zi was himself a disgruntled Confucian, wishing only that Confucians would live up to Confucius' ideals, which surely were supposed to be universal. When one of Confucius' disciples Ji Kang Zi, destined to rule a third of Lu (all Lu having been divided into three parts), came bounding up with a brainstorm about good government, "How would it be if I killed those who oppose the Way for the sake of those who uphold it," Confucius replied in a six syllable counter-question that remains one of the best ever asked: *"Zi wei zheng, yan yong sha?"* (In governing, why must you kill?)[67] But Confucius himself did not think governors could stop killing over night. Elsewhere he said: "True indeed is this saying: 'If good people ruled for a hundred years, they could overcome cruelty and do away with killing.'"[68] After they had killed off those who opposed the Tao?

Confucius sounded like Lu Xun: "The good people will ask, perhaps, 'Then, do we not want fair play at all?' I can immediately reply, 'Of course we want it, but it is still too early'"[69] (an unwitting echo of Augustine: "Make me chaste and continent, but not yet."[70]).

Mo Zi's universal or inclusive love was not inclusive enough to include "violent people" *(baoluan zhi ren)*, usually taken to mean "rebels." Confucius thought the death penalty would be necessary, even under the best of conditions, for the better (or worse) part of a century. In Moses' camp, God himself was said to have ordained the death penalty for seven specific sins, in his oral footnotes to the ten commandments.[71] So from the very beginning, in cultures round the world, "Thou shalt not kill" has not been taken to mean, "Thou shalt not execute."

Lu Xun was for the death penalty. Less (or more) sanguine than Confucius, he thought it would be needed even in a golden age.[72] But we have never seen a golden age or a Confucian century. What counts is simply that Lu Xun took his place with Moses, Mo Zi, and Confucius among the sublime and ridiculous billions, from the God of the Old Testament to Msrs. Reagan, Bush, and (alas) Clinton, who have declared it right on occasion to beat dogs to death in the water.

In accepting Chairman Mao's notion of "common sense," before it was Chairman Mao's notion of common sense, Lu Xun had plenty of company.

So where is the "great contradiction?" It is where it has always been, staring us straight in the face. However common the practice, there *is* a contradiction, on the face of it, between opposing people-eating and eating people-eaters—if people-eaters are people. If they are not, then and only then does "the great contradiction" vanish like a mirage. So that is the question, a philosophical question, perhaps a scientific question, *not,* nonetheless, an academic question: Are people-eaters people?

Darwin told Lu Xun they are not.

All through history, of course, people have found it perfectly easy to believe they are not, without the slightest help from Charles Darwin. Even the "religious" have told us that people-eaters are not people. People-eaters have been called demons, witches, agents of the devil, "the forces of darkness."

Lu Xun too used such language. If the "present state of chaos" was not to last forever, then Light," he said, would have to "wage all out war with Darkness."[73] That is why he decried showing "mercy to demons and devils,"[74] (just as Mao Zedong would, promising us, as he did, as late as 1964, that "all demons and devils shall be annihilated"[75]). Granted, such language was hyperbole. Lu Xun did not believe in the "good old-time religion." And yet he did believe in his Haeckelian Darwinism, and that led to the same result: to the dehumanization of the opposition.

Bad science took over from bad religion.

"Alas," growled Lu Xun, in a letter written in 1935 to Xiao Jun and Xiao Hung, "why are there so many human-faced dogs!?"[76] Hyperbole again—he was venting his indignation against Guomindang censors. But that curse describes all too literally the "people-eaters" he denounced in 1918, the "dogs" he said should be beaten, even in the water. For "people-eaters" looked like people but were not. They lived among us, but were a breed apart. They came in all colors. They did not belong to only one of the laughably labeled human races: red, yellow, black or white. But human-faced, of many hues, they together were a race, fixed by nature, by a nature that revealed itself not in visage, although sometimes people-eaters had wolf-like eyes,[77] but in behavior, behavior that would not soon change. That was why one should show them no mercy, because "dog nature is not very susceptible to change. Perhaps after ten-thousand years it might be different from now, but what I am talking about now is now."[78]

Why was that bad science? Because there is no "direct evidence" whatever—this time I applaud Gould's phrase—"for genetic control of specific human social behavior" of the sort that Lu Xun thought distinguished people-eaters from people.[79]

"Bad science," of course, is a stone that even good scientists should throw with caution. I throw it with fear and trembling, admitting that Lu Xun's position was not totally unscientific. I think it vital for readers of Lu Xun to see that Lu Xun leapt from scientific fact to unscientific conclusions, but even though he did so, he landed (as my metaphor stretches almost to the breaking point) with one foot still in evolutionary science. That was his position in his day. That is still his position in ours.

What were his grounds for believing that "human-faced dogs" might dwell among us? He accepted the evolutionary fact that our ancestors have not been of one nature. The nature of our ancestors has been in flux. It does not follow, however, that *human* nature has been in flux, however variable human genetic recombination and mutation, for "human nature" is a label that can be pinned on something only for as long as it does *not* seem to vary significantly. Human nature did not evolve out of human nature. It evolved out of protohuman nature, and if anything evolves out of human nature, it will be posthuman nature. (*Nota bene* once again: proto and post make scientific sense; sub and super do not.) Human nature, therefore, refers to the nature of the beast that we see before us—in space and in that little bit of time that we call history, in which we have not yet found any meaningful change in the nature of our species, not, at least, in the nature of *homo sapiens sapiens*—seeing, wherever we look in the mirror of history, ourselves.

However, have not new species evolved from old species? Of course. Have not new species "surfaced" from mixed populations? Well sort of. Might not "human-faced dogs" and human beings exist then at the same time? Yes, but if they could interbreed productively they would still be of a species. But species can have recognizably different varieties? Of course. So a *race* of human-faced dogs or people-eating people might exist? *Conceivably,* but we would have to look for one.

"Races" exist, many more than we usually recognize. People quite arbitrarily recognize as races groups distinguishable (not always easily) by color. We might just as well recognize races of the multihued tall, or short, of multihued people susceptible to certain diseases, of multihued painters, musicians or mathematicians. Most important, perhaps, we might recognize multihued races of the attractive and the unattractive, as determined by success in reproduction.

Do "races" of musicians and mathematicians pull us from physical characteristics to behavioral characteristics? We are not sure of the difference. Do we admit that "races" might be recognized by behavior? *Yes.* Even without accepting absolutely the sociobiologists' creed that there is a "biological basis" for "all social behavior," we accept the notion of behavioral races when we accept scientists' claims that there is a biological basis to the multihued "race" of manic-depressives, for example.

Then why could there not be a race of people-eaters. There might be. It is not totally unscientific to imagine that there might be. But can we find one? *Maybe.* Some of the homicidal people dismissed from our courts as "not guilty on grounds of insanity," only to be caged (like human-faced dogs?) in insane asylums, *might* be members of such a race. But even at its most inclusive that race is far too small to be the race Lu Xun thought should be shown no mercy.

Lu Xun never said that "people eaters" were the "congenitally insane." He was inconsistent. He pinned his "people-eater" label on two different groups, but never on that group. In "A Madman's Diary" he pinned it on us, on all people, except, perhaps, for a few children. In "On the Advisability of Postponing Fair Play" he pinned it on "them" on a vast group of "human-faced dogs," who were "those opposed to reform."[80]

That is the group we must hunt for. But what "direct evidence" is there that there was or is in China a biologically determined race of people-eaters opposed to reform? "At the moment," to borrow Gould's words one last time, "the answer is none whatever."[81] For even real people-eaters have quit. *On occasion* cannibals have become ex-cannibals, human carnivores have become vegetarians, alcoholics have become teetotalers. Murderers have stopped murdering, torturers have stopped torturing, exploiters have stopped exploiting. With or without local PA chapters (People-eaters Anonymous) some people-eaters have stopped eating people. And here it makes no difference whether they stopped through choice or conditioning, for in either case the point is the same: The specific behavior that Lu Xun loathed was not irrevocably determined by nature—and it was bad science to say that it was.

Lu Xun's Haeckelian Darwinism led him, on occasion, to take the position of a behavioral racist.

I find that position fraught with sad ironies and with even sadder logical, and practical, consequences.

Lu Xun's "evolutionary" argument echoed, and thus ironically supported, the "evolutionary" arguments of ordinary nineteenth-, and twentieth-, century Social Darwinian racists, one of whom he was not.

His behavioral racist arguments against imperialists echoed, as we have seen, and so ironically supported, the racist arguments of imperialists, whom Lu Xun detested.

In his argument that "fair play" must be postponed because "the human race has not yet grown up, and the humanitarian way [rendao] has naturally not yet grown up either,"[82] Lu Xun, in so many ways like Huxley, sounded instead like Herbert Spencer. Spencer believed that the human race was growing up, and in growing up was becoming more humane. "Sympathy," he said was "the faculty which distinguishes the social man from the savage."[83] But he savagely denounced the welfare system of his day, because it extended sympathy to "the unfit," the poor, and thus retarded "the process of adaptation," the "purifying process" that in time would produce "the ideal man."[84] It was too early for universal sympathy—or fair play.

Lu Xun was even tougher than Spencer, because the evolutionary low life he feared were people-eaters not just poor people. Spencer said of the poor, "If they are not sufficiently complete to live, they die, and it is best they should die."[85] Lu Xun said of "those who oppose reform," "If there are people who want to extend the spirit of 'fair play' generally, I think they should at least wait until 'dogs in the water' show some signs of humanity.... We must make distinctions. Before being 'fair,' we must look at whom we are up against. No matter how they fall in the water, if they are human beings help them, if they are dogs let them be, if they are bad dogs, beat them."[86]

Both Lu Xun and Spencer were against what Spencer called "unthinking benevolence."[87] In the name of sympathy, Spencer railed against "sigh-wise and groan-foolish" people "disabled ... by their sympathies."[88] In the name of *rendao* (humanity), Lu Xun railed against *rendaozhuyizhe* (humanitarians), who "let snakes return to their ravines"[89] (to give the dogs in the water a rest). In 1925 Lu Xun was thinking of those who showed "unthinking benevolence" to people who opposed the Revolution of 1911: "True hearted, fair minded people started shouting, 'Do not take revenge, be compassionate, do not resist evil with evil....'"[90] In 1935 he was still thinking of them, and of others like them: "Twenty four years ago we were too magnanimous. We were fooled by the word 'civilized.' In the future, too, humanitarians will oppose taking revenge. I hate them."[91]

What justified this "toughness?" A Spencerian view of evolution, a Spencerian conviction that we should follow the Way of Nature: "Pervading all Nature we may see at work a stern discipline which is a little cruel that it may be very kind."[92] Or a Spencerian interpretataion of Lao Zi's words: "Heaven and Earth are not humane ... the sage is not humane"[93] (or the "superman" or the "extraordinary man?"[94]).

There was a contradiction in Lu Xun's logic: and he did not deny it, he embraced it:

If we want "fair play," we had first best be sure of our opponents. If they do not deserve "fair play," we quite frankly need not be polite.

It will not be too late if we talk with them of "fair play" after they "play fair."

I may well seem guilty in this of advocating a double standard, but I do so of necessity. Otherwise China will never be able to find any relatively good road.[95]

So this was his conclusion:

I dare state categorically that the counter-reformers have never let up in their poisonous attacks on the reformers. Indeed there is already no way they could be fiercer in their methods. The reformers are the ones still dreaming, the ones who always get the worst of it. That is why in China there is never any reform. Hereafter we should change somewhat our attitude and our tactics.[96]

But what did he really mean? Was he thinking of "impolite" pen wars or real wars? Was he calling writers to arms, or students, or soldiers? And to what arms did he call them, words, or sticks and stones? He did not say, but his *language*, in "On the Advisability of Postponing 'Fair Play,'" was militant. The phrase with which he professed to sum up what he meant was militant. Again, he did not deny, he embraced an accusation from one of his opponents (just as Mao Zedong would embrace that famous accusation from his opponents: "'You are inhumane'"):[97]

Before being fair, we must look at whom we are up against. No matter how they fall into the water, if they are human beings help them, if they are dogs, let them be, if they are bad dogs, beat them. *In a word "dang tong fa yi"* (unite with like, attack unlike).[98]

What did he mean by "attack?" Again, he did not say, but his words could, and can, stand as an invitation to violence.

Eight months earlier, to the day, he had already used violent language to call for a change of tactics, reverting to his cannibalism metaphor to decry the inhumanity of his "civilization":

"Chinese civilization" is actually only a feast of human flesh, arranged for the enjoyment of the rich. "China" is actually only the kitchen where this feast of human flesh is prepared. If people extol it out of ignorance, they may be forgiven. Otherwise they should be damned for all eternity.

This civilization has not only intoxicated foreigners, it has long since intoxicated all Chinese—so that they smile. Because all have been isolated by various distinctions handed down from ancient times, and still present, none can sense the suffering of others; and because all have hopes of enslaving and eating others, all forget that they themselves shall be enslaved and eaten, in the future. So large and small, innumerable feasts of human flesh have been set, from the dawn of civilization until today, and people in this hall have gone on eating people and being eaten, the imbecilic shouts of the savage drowning out the miserable shouts of the weak, not to mention those of women and children.

This feast is still set, and many still want to keep it set. So the mission of the young today is to wipe out these people-eaters, to break up this feast, and to destroy this kitchen![99]

A far cry from the madman's cry: "You can change." This was a call not for repentance but for a purge. Lu Xun had changed his audience. He was no longer calling on his people, people-eaters all, to change their evil ways. He was sicking young people on nonpeople, casting out people-eaters as a race apart, as a race beyond redemption. He was, *rhetorically, at least,* calling for a holy war, against *some* of his people. Whatever he had in mind, he had put in print a call for civil strife—arguing, of course, that it would not be civil strife because people-eaters were not of the people, a tenet-to-be of the Thought of Mao Zedong, that "enemies of the people" were not of the people.

That was a radical change of position for Lu Xun. But did it, as the orthodox, or those constrained by orthodoxy, have consistently maintained in the People's Republic, represent a great leap forward in the development or evolution of Lu Xun's thought? No. The cries "You can change" and "wipe out these people-eaters" are poles apart, but they represent a lasting polarity in Lu Xun's thought not two points in any linear evolution—because Lu Xun remained torn, to the last, between the two.

Early and late his message was mixed. The madman's message was mixed. He told people-eaters they could change, but he also told them they would be exterminated if they did not. But the madman *"recovered."* Did that mean that people-eaters could not change? "Save the children," he said. But *who* could save the children? Where could "the young" come from who could "wipe out these people eaters?" Lu Xun began that very passage insisting that "all Chinese" were intoxicated with China's cannibalistic civilization. Who, then could destroy it? Was he calling down the wrath of "evolution" on his people? But in the same year he said, "The next reform is to have the people reform their own evil nature," as if they could. So sometimes the enemy was "us," sometimes it was "them." Sometimes he said, "Purge our land of sinners," sometimes he said, "Purge ourselves of sin." Sometimes he said, "We must stop eating people," sometimes he said, "We must wipe out these people-eaters." He did not switch once and for all from one cry to the other. He went back and forth. He could not make up his mind.

But his words could help others make up their minds—and they did, or at least they "justified" others, whose minds were already made up. And they still can.

And some of his words were violent. His Haeckelian theory of dogs and men, his "Darwinian" double standard, was violent and invited violence, or "justified" violence post facto—violence that Lu Xun did not have in mind.

To whom did he first offer his Haeckelian theory, his double standard, his militant advice, to "join with like and attack unlike?" To the Nationalists. In 1926 he wrote to Xu Guangping: "The Nationalists are too naive. . . . They are magnanimous to members of other parties who are not magnanimous to them."[100]

And what did the Nationalists do? They did not, of course, wait for Lu Xun to tell them what to do, but on April 4, 1927, four days after Lu Xun gave a speech to cadets at the Whampoa Military Academy, in which he modestly minimized the importance of "revolutionary literature," protesting that he would rather hear from fighters than writers,[101] the Nationalists did indeed decide no longer to be "magnanimous." They jumped the members of the major "other party," their allies, the Communists, and slaughtered them like dogs in the water.

Six years later, when Lu Xun finally published his letters to Xu Guangping (and hers to him), he deleted the line about the Nationalists' magnanimity.

How many more lines would he have deleted had he known to what use they would be put?

In the last years of his life, Lu Xun offered his militant rhetoric to the Communists, and they accepted it with thanks. But Lu Xun did not live to see what they did with it.

On the first anniversary of Lu Xun's death, October 19, 1937, Mao Zedong himself gave a speech in memory of Lu Xun, in which he urged Chinese to "study the spirit of Lu Xun":

> In one of his essays he advocated the beating of dogs who have fallen in the water. He said: If you do not beat dogs who have fallen in the water, once they jump out, they will not only bite you but splat-

ter you, at very least, with foul muck. So he was for finishing them off. He made no hypocritical show of being "merciful." At present we have not yet beaten the mad dog of Japanese imperialism into the water. We must keep beating that dog until it has no way to recover and must retreat from China. We must study this spirit of Lu Xun and employ it throughout all China.[102]

And Mao Zedong did, against the Japanese, against the Nationalists, and then against "class enemies."

Let us not dwell on the Japanese and the Nationalists, although they too were people, not dogs, and even in beating *them* in the water, people could eat people. Granted, the "feasts of human flesh" spread in China by the Japanese and the Nationalists, though not comparable, were each so revolting that only an absolute pacifist, and we know Lu Xun was not one, could question absolutely Lu Xun's question, "Why *not* struggle?," or the Chairman's dictum, "To rebel is justified." Still, even in resisting "wolves and tigers"[103] there was danger. Eating one's enemies, one could acquire their tastes. In the long course of the civil war, only somewhat interrupted by the war against Japan, the Communists became "blood brothers" with the Nationalists.

But the saddest feast would come later. For decade after decade *after* the Communist revolution, Lu Xun's Haeckelian theory of men and dogs would be used to support the Chairman's theory of class struggle, one Social Darwinian theory supporting another.

Marx said Darwin discovered a "basis in natural science for the class struggle in history."[104] Mao Zedong said, "Class struggle, some classes triumphing, some classes being eliminated, that is history"[105]—natural history. Classes struggled to survive, some fit some unfit, fit and unfit, indeed, in turn, representing, like Haeckel's races, different stages of evolution. Men of vision, like Haeckel and Marx, could foretell the outcome of such struggle. And yet, as the Chairman said, "reactionary classes," evolutionary low life, although doomed, would never "take themselves off the stage of history," for "things reactionary will not fall unless you hit them."[106]

Individuals belonged to classes. They shared no "abstract human nature." They were divided by "class nature," which, like "dog nature," was not "very susceptible to change." It is true the Chairman wavered on this point. Sometimes he suggested that reactionaries could change their spots. But over and over again, in his people-eating periods, he classified people according to the classes of their fathers—or grandfathers, as if "class nature" were hereditary. So capitalists' grandsons who worked in factories were capitalists, and landlords' grandsons who worked the land were landlords, and bourgeois intellectual grandsons who had never been to high school were bourgeois intellectuals, or as the risible jingle of the Cultural Revolution put it, in the ultimate expression of this natural law: *"Laozi yingxiong er haohan, laozi fandong er hundan"* (If the father's a hero, the son is brave, if the father's reactionary, the son's a knave)—risible now, at the time no joke.

If thought and deed are determined as Marx said they were, by how we are given, or get, our daily rice, then there can be no "basis in natural science" for the Chairman's theory of class nature—*unless*, as Herrnstein suggests they might, our *genes* can determine our class.[107] Marx, however, did not believe in genetic determination of class. He was a behaviorist: change the mode of production and you can change men's consciousness. Change a man's rice bowl and you can change a man's soul. But if that is so, no one should be an irredeemable exploiter, or people-eater. So at least *after* seizing power, why break heads instead of rice bowls?

Mao Zedong too, in his darker moments, was a "racist." He divided his people into "the people" and "enemies of the people," dehumanizing "class enemies" and "counter-revolutionaries" as Lu Xun dehumanized—rhetorically, at least—"counter-reformers."

What came of the division? Again, a double standard, grasped perfectly by Mao Zedong's model student, Lei Feng—"Be merciless to your class enemies. Be warm as spring towards your comrades,"[108] a pervasive double standard still visible in a dictionary definition published in 1979:

Cibei mercy; benevolence; pity: *fa cibei* have pity; be merciful/*Dui diren de cibei jiushi dui renmin de canren.* Kindness to the enemy means cruelty to the people.[109]

What came of that double standard? People eating.

Granted, class *can* influence consciousness and action. Granted, in "the old China," capitalists, landlords, warlords, and Nationalists all too often did eat people. Granted, the goal of the revolution was to create a world in which people would not eat people. Granted, the government of the People's Republic of China has done more to serve the people than any government in Chinese history. Nonetheless, the government of the People's Republic of China has also eaten more people than any government in Chinese history, in all its great political movements, the suppression of Counter Revolutionaries, Land Reform, Thought Reform, the Anti-Rightist Movement, the Great Leap Forward, the Great Proletarian Cultural Revolution, in its general suppression of all "enemies of the people," real or imagined, and in its often inordinate demands for self-sacrifice from even the most ordinary of "the people," in its expectation for example, that millions of husbands and wives should willingly serve the people separately, in different towns or even provinces.

Too much of all that suffering—after the revolution—came from a paranoia born of Social Darwinian-Marxist superstition, a paranoia born of the superstitious belief that "the people" were engaged perforce, per an historical, evolutionary force, in a struggle for survival with a dying but diehard subhuman species of enemies made enemies by nature.

Lu Xun supported that superstition, never recognizing it for what it was. And so he must share responsibility for the suffering that came of it. Lu Xun's first version of that all too pervasive superstition was Haeckelian not Marxist. Lu Xun first offered it to the Nationalists not to the Communists. Indeed, the first time he added any Marxism-Leninism to his theory, he was still offering it to the Nationalists not to the Communists. Only *two* days before the Nationalists first tried to annihilate the Communists, Lu Xun urged them to take *Lenin's* advice to victorious revolutionaries: "Annihilate the enemy."[110]

Jiang Jieshi tried to do just that. He did not actually take Lu Xun's advice—or Lenin's! He "instinctively" embraced the great superstition, but he *could have* quoted Lu Xun's essay "On the Advisability of Postponing 'Fair Play'" to justify the terror of his "party purge."

Lu Xun never faced that fact—not even when he deleted that line about the Nationalists being too magnanimous to their enemies.

In September of that horrible year 1927, Lu Xun wrote:

I once said China has played host to one continuous cannibalistic banquet. Some have eaten, some have been eaten. Those eaten have eaten. Those eating shall be eaten. But now I have realized that I myself am helping to host that banquet.[111]

But he did not realize *how* he was helping, or how he *would* help. He only saw that he was setting people up to be eaten. He did not see that he was encouraging other people to eat them.

Once he almost saw it. In one clairvoyant, sarcastic moment in 1933 he wrote:

China is indeed the oldest seat of civilization, a country that has habitually honored the humanitarian way. Human beings have all along been shown the utmost respect. If some, on occasion, have been humbled or killed, it is because the things were not human.[112]

But he forgot that he had said the same thing, that counterreformers were not human—and so deserved no mercy.

Two years earlier, sickened and infuriated by the Nationalists' arrest, imprisonment and secret execution of five leftist writers, including his friend and protégé, Rou Shi, Lu Xun said that China's rulers had thereby proven that they were "animals of the dark, on their way to extinction."[113] That was righteous indignation and "perfectly natural" hyperbole, but the "theory" behind it would "justify" similar outrages yet to come.

Lu Xun did not see the danger, that those who dehumanize their enemies dehumanize themselves. Lu Xun witnessed the people-eating of the warlords and the Nationalists. He knew of people-eating in every class of the old society. And he said the worst people-eaters should be beaten—even in the water. But he did not live to see the beating, or eating, his words would "justify."

He could not foresee that as early as 1942 his own disciples would be beaten—in his name.

He could not foresee that over the course of thirty years, millions of people would be beaten in the water.

He could not foresee, of course, the Cultural Revolution, nor, after it, the sad report of a leading Lu Xun scholar:

> Because the "Gang of Four" distorted Lu Xun, quoted his words out of context and used them as clubs to beat people, many young people who grew up in the Decade of Disaster, not understanding very well the true meaning of Lu Xun's works, have come to look on them with repugnance.[114]

This is the ultimate irony. Some of "the children," somehow "saved," with Lu Xun's help, perhaps, but more immediately by the pain of their own experience, people who were children when he wrote his stories, or later, children of the People's Republic, sons and daughters of the revolution, people revolted by people-eating in the People's Republic, who have finally revolted against people-eating in the People's Republic, modern "madmen" who have finally attacked people-eating even in the form of "socialist alienation," calling for *rendaozhuyi* (humanism or humanitarianism or humaneness), some of such people have put Lu Xun among the people-eaters, as if he himself was a "madman" who "recovered."[115]

Surely that is the saddest of ironies, but is it a gross injustice? Not absolutely. Granted, the "Gang of Four" could distort things. Granted, too, Lu Xun would not have liked the Gang. But the Gang could honestly like Lu Xun (dead not alive), for he did indeed lend himself to their cause.

For their cause in the Cultural Revolution, the Gang of Four were quite willing to distort works of history and literature. Their scholarship, or the scholarship of their scholars, semicaptives like the "Liang Xiao" group,[116] was academically dishonest. Their studies of Confucius, Zhou Gong, Qin Shi Huang, Wu Zetian, and the *Shui hu zhuan* were cynical fabrications written to attack the persons and policies of Zhou Enlai and Deng Xiaoping. For their cause, when they felt the need, they were perfectly happy to "distort Lu Xun."

They did "use" Lu Xun, and Lu Xun would not have liked their use of him. Lu Xun would not, I suspect, have liked "the Party's" use of him at many important junctures in the history of the People's Republic, and had he been around to protest, "the Party" might have liked him much less than it liked the useful, albeit always somewhat dangerous, legacy of his works.

Such speculation is not idle, for it has been and remains more than an intellectual exercise for Chinese intellectuals to wonder how Lu Xun would have met the challenges they themselves have been forced to meet. What would he have done when his disciples were attacked in 1942? What would he have done during Thought Reform, The Hundred Flowers Movement, the Anti-Rightist Movement, the Cultural Revolution? And what would have been done to him?

One brave, or foolhardy, citizen of Shaoxing, Lu Xun's hometown, wrote in 1980, in a poem published in *The People's Daily*, entitled "If He Were Still Living" (if he had lived to be ninety-nine!):

> Perhaps he would already have received all sorts of honors.
> But perhaps—he would just be getting out of jail.[117]

Had he lived, it is likely that Party leaders, from the Chairman to Liu Shaoqi (maybe even to Zhou Enlai), to the Gang, to Deng Xiaoping—all might, in time, have found him hard to take. But dead, he could help them all, and he did. So the Gang perhaps liked him honestly. They did not have to distort him or misquote him or quote him very much out of context. They had only to quote him in his anger. For Lu Xun in his anger, in his "righteous indignation," could indeed support violence, vengeance and ruthlessness.

Lu Xun supported these things *in theory*. In practice, he supported violence only three times, and then only in the most general terms. He supported all three of China's revolutions, but always from the sidelines, never loudly, and never with infectious conviction.

As we have seen, he supported the Revolution of 1911, but more obviously after it than before it. Before it, he neither took part in nor praised assassination attempts or uprisings.

He supported the Nationalist's Northern Expedition. In the middle of it, as we have seen, he told cadets at the Huangpu Military Academy that he longed to "hear the sound of cannon."[118] And a year before it, in an "inflammatory" statement, quoted since, ten thousand times, in praise of the Communist Revolution, he said "the quickest way to reform is through fire and sword."[119] But he said that privately to Xu Guangping, in a letter not published for eight years. Moreover, when Xu Guangping, thrilling to Lu Xun's silent call to arms, suggested in reply that heroes should "attack with three-inch swords [sic] traitors who oppose the people's will, and then should shout their defiance to the skies, fall upon their swords, and die," Lu Xun hastily wrote back, "I really have no way to say the suggestion in your letter is definitely wrong, but I do not approve,"— not because he had moral qualms about the assassination of warlords ("It would make people happy for awhile"), but because "it would not affect the greater situation," and because he did not like urging others to take risks that he had not taken—and was not about to take.[120] His praise for fire and sword remained academic.

After the Nationalists' "Party Purge" of 1927, Lu Xun transferred his sympathies to the Communists. He probably looked forward to a Communist revolution. But in print (he wrote, of course, under severe Nationalist censorship) his one open declaration of support for the Communists in the field, and his one open declaration of support for Mao Zedong, was actually only a declaration of support, on the face of it, for Mao Zedong's call for a united front against the Japanese. "I deem it an honor," Lu Xun wrote, "to be able to call comrades those who are actually there with their feet upon the ground shedding their blood, struggling for the present survival of the Chinese people."[121] That was only an echo of the earliest Chinese use of Darwinian idiom, an echo of the ultimate cause of Yan Fu, Liang Qichao, and Sun Yatsen—"the survival of the Chinese people." It was no call for the annihilation of "class enemies." Lu Xun never, not even in private correspondence, specifically urged the Communists ruthlessly to suppress "class enemies" or "reactionary intellectuals." All he did was give them a pseudoscientific justification for such suppression. But he did give them that, and they used it.

And why not? Had he lived to decry their use of his law, what could he have said? Only that they had answered erroneously the Chairman's question, "Who are our enemies? Who are our friends?"[122] He could not have said they were wrong to eat their enemies, for all creatures had to— to survive. He could only have said that they had misidentified friends as enemies, or that *they* were enemies, people-eating dogs who themselves should be beaten in the water. But if he said that, the beating—and eating—would go on: The vicious cycle and the same dilemma. "If he were still living," what would he say, "Let us stop eating people" or "Beat the people-eaters?" Or would he still say there is no contradiction?

He died before he saw that his law was flawed.

Lu Xun said that human beings, "true human beings," would be humane. He also said that it would not be inhumane to be inhumane to the inhumane, because the inhumane were not human. He was therefore at once for humanity and "inhumanity." And he used Darwin in defense of both.

Darwin, he said, said that "true human beings" were evolving. Darwin, he said, said that people-eaters, although they looked like people, were not people.

Darwin actually said nothing of the sort. Darwin never said we could know where evolution was going. Darwin never said that a species could "distinguish" itself by ideal behavior. Darwin never said that people who looked like people and could successfully breed with any people who looked like people, could not be people. Present people were always of one species.

Can we "blame" Lu Xun's double standard, then, on his Haeckelian Darwinism? We still do not know enough about thought to know. Haeckelian Darwinism helped convince him or helped confirm him in his conviction that it is wrong to eat people but right to postpone fair play. But beneath *both* sides of that conviction we can see Confucianism. Humanity was the heart of Confucianism. Humanity defined humanity: *"ren zhe ren ye"* (the humane is the human).[123] Over and over we encounter that homonym: *"Fan Chi wen ren. Zi yue 'Ai ren'"* (Fan Chi asked what humanity meant, and the Master said "Love humanity,")[124] all humanity, for "within the four seas all are brothers"—and sisters! Despite that last lapse, Confucius sounded as if he had only one standard: "Confucius said," said Mencius, "that there are only two ways, the humane and the inhumane."[125]

But *Confucians* had always had a double standard. For two thousand years a Legalist-Confucian compromise had been the rule, all too often, at least, for official and family Confucianism. This was what Professor Schwartz long ago called "muscular Confucianism,"[126] which had shown about as much mercy toward evil as had the medieval Christian church. Lu Xun understood this unforgiving "righteousness." He understood how easily it could eat people, evil and innocent, in the tangled mess of human affairs. That is why he said *renyi daode* (humanity, justice, and morality) ate people. But he did not see that the harshness of his own *zhengyigan*, his sense of justice, came from the same tradition. He hated that tradition. He was afraid that it corrupted all Chinese. But he did not see how it led him to prolong the feast himself, even as he called for the destruction of the kitchen.

When he said "We must stop eating people," he stood with Confucius. When he said "We must beat dogs in the water," he stepped instead, without realizing it, into the ranks of the "muscular Confucians."[127] He rationalized his "muscular" Confucianism with "muscular" Darwinism.

But was that ruthless "righteousness" simply the joint product of "bad" Confucianism and "bad" Darwinism?

No. Of course not. Lu Xun's reasoning came second. First came an elemental hatred of evil and an "instinct" (almost—we do not understand it) to strike out against evil, even, if necessary, with the weapons of evil.

He knew, of course, that moralists warned us against doing that, against fighting fire with fire, against "exchanging violence for violence."[128] He knew that Confucius said we should only fight evil with good. He knew that Mencius promised that good would triumph. For "humanity conquers inhumanity," Mencius said, "as water conquers fire." Those who do not believe it are "like those who try to put out a burning cart full of fire wood with a cup of water and, when they fail, then say that water does not conquer fire."[129] But Mencius himself had not always kept that faith. At least once he seemed to justify tyrannicide. Twice he seemed to say, two millennia before the Chairman did, that, sometimes, "to rebel is justified."[130] And even Confucius said, as we have seen, that a little muscle might be necessary to suppress evil—for the first hundred years.[131] Alas, he was optimistic. Muscular Confucians have not gotten beyond the "Dieu et mon droit" school in the first twenty-five hundred years.

Small wonder that Lu Xun could not believe that right was might enough. Everywhere he looked he saw, as the proto-Darwinian idiom put it, that "the weak are the meat of the strong."[132] And he thought the weak should resist. So what he said of Byron's angry sympathy for enslaved peoples was true of his own angry sympathy for his own people: "He grieved at their misfortune. . . . He raged at their refusal to resist."[133]

And he came to loathe, therefore, "true hearted apostles of reason" who "took up the shout," when the oppressed did resist, "'Do not seek vengeance,' 'Be compassionate and forgiving,' 'Do not resist evil with evil.'" For the loudest of such people, he thought, were running dogs of the people-eaters, fed at least on scraps of human flesh. That is why he said, of "humanitarians who oppose taking vengeance," "I hate them."[134]

But however "understandable" his anger, however "reasonable" his hatred of "true hearted apostles of reason," and however cannibalistic Nationalist rule, what was bound to come of his Darwinesque dog-eat-dog call to eat people-eaters? People eating, as surely as vengeance comes of vengeance, the *lex talionis* of the ancients.

And that is what did come of it. The Communists ate their enemies, and then, in their fear, seeing dogs in every pond, they ate their own. Their people and each other. The Chairman indeed became a cannibal.

For fifty years, Lu Xun "officially" supported him. Dead, he who first raised the madman's cry against people-eating, "officially" supported people-eating. And scholars and officials who in our day say, "Nonsense," who protest that "dog-beating" is not people-eating, have supported the new people-eaters and have themselves fed on scraps of human flesh.

Lu Xun, blinded by righteous indignation, gave the Communists "good reason" for ruthlessness: Their enemies, he said, were inhuman. He also gave all people the simplest and best reason against it: People should not eat people.

"If anyone has ears to hear, let him hear."[135] For the last fifty years who *has* heard what, is history. What counts, is who *will* hear what.

Both cries, "Do not eat people" and "Beat dogs in the water" are still there, both powerful because Lu Xun's works are still powerful. The "great contradiction" is still there, because the great dilemma that it reflects is still there. The ancient question is still there: Is there a good way to fight evil?

Confucius said, "There are only two ways, the humane and the inhumane."

But was there a humane way to take power from the Nationalists, who had become disgustingly inhumane?

The Communists rejected the humane way, which seemed to them no way. That is why they have officially continued to reject *rendaozhuyi* (humanism or humanitarianism), as a counter-revolutionary philosophy. They must support revolution, their raison d'être.

But what now? Since the Cultural Revolution, Chinese have called the *Communists* disgustingly inhumane. They have called the Communists people-eaters.

The Communists have turned daringly, and honestly, toward reform, "for the people,"[136] for the country, and for themselves, "for themselves," because surely they have heard on all sides, at least faintly, "the songs of Ch'u."[137]

In their struggle to survive, will "fair play" be postponed further? Or will they remember the madman's warning, "You must know that the future will not allow people who eat people to live on this earth. If you do not change, you yourselves will be devoured"[138]—the warning of Paul, "If you bite and eat each other, watch that you be not consumed by one another."[139]

In this current ferment, officials and the disenchanted both tell people, on occasion, to "study Lu Xun." But study what? What, if they actually listened to Lu Xun, would they hear? Would officials hear "You can change," or "Beat dogs in the water?" Would the disenchanted hear "To rebel is justified," or "eating people is wrong?"

Will bloody struggle come again? Will the *lex talionis* go round again, or "the dialectic?" Or will Chinese somehow tame the Furies?

The way is not clear. The dilemma is still there. But who can look at China's century of civil strife and wish for more? Who would not say, "Amen" to the prayer of the leader of the Furies, that ancient prayer voiced thanks to Athena's magic—which still escapes the world—the magic that transformed the Furies into the Eumenides:

> I pray that civil strife,
> Which knows no end of evil,
> Shall never roar within this city.
> And may the dust
> That drinks the black blood of its people
> Wreak not havoc on the state,
> In rage demanding recompense
> Life for life.[140]

But Furies and Eumenides warred in Lu Xun's rhetoric to the end. In the last line of the last essay he ever wrote, Lu Xun said:

We have been born with human heads. Let us speak like human beings.[141]

In a letter a few days later, he said of the author of the essay he attacked in his (and he attacked it with good reason):

Truly, this author is a swine.[142]

9

The Evolution of Lu Xun

The title of this final chapter is ambiguous—to cover our two remaining tasks: a final evaluation of Lu Xun's use, and understanding, of Darwin's theory of evolution, and a final evaluation of the contention of Mainland scholars that the truly important story, in the story of Lu Xun and evolution, is, indeed, the story of *his* evolution from a patriot with an ill-founded faith in evolution alone, to a patriot with a well-founded faith in the revolutionary evolutionary science of Marxism-Leninism.

Two final efforts—the one of which will help the other:

Pseudoscience to the Rescue

Lu Xun wrote no more essays about the theory of evolution after writing his "History of Man" in 1907. He *used* the theory of evolution, however, in dozens of essays about other things. Once he started trying to wake people up in "the iron room," he used Darwinian language again and again, angrily, cleverly, humorously, sardonically—almost always effectively, but almost never scientifically.

I say this not to denigrate Lu Xun, but to warn those who read his works, and works about him, not to *zi qi qi ren* (to fool themselves and fool others), as so many have in the People's Republic. Lu Xun's greatness does not lie in his grasp of science, natural or social.

All his life he was "for" science. As a young man he studied geology and medicine, and dabbled in botany. Of the two messiahs of the May Fourth Movement, he had much more faith in "Mr. Science" than in "Mr. Democracy."[1] He never professed faith in democracy. Early and late he had too low an opinion of "the people." In his early essays he cast doubt on the wisdom of majority rule: "If a man lives with a mob of monkeys, must he live in the trees and eat acorns [*sic*]?"[2] At the height of the May Fourth furor he still decried "mob"[3] rule: "Men of foresight have always been suppressed, rejected, ruined, exiled and killed by vicious, petty people and the muddleheaded masses."[4] It was "Mr. Science" who would be China's savior if Chinese would receive him: "I have always hoped that the damage from our heredity of muddleheaded thought would not be as severe as that of syphilis, which spares none of those who get it, but even if it were, 606 having been invented, that physical illness can be cured; I hope there is also some 707 medicine that can cure the illness in our thought. —But such a medicine has already been invented. It is 'science.'"[5]

The trouble was that the masses were too muddleheaded to take their medicine:

> "Save the nation with science"—has been shouted for nearly a decade, and everyone knows that cry is correct. . . . [But] every new system, new branch of learning, new term, that reaches China, falls, as it were, into a vat of black dye, becoming immediately pitch black, transformed into a tool for selfish evil. —Science is but one example.

> If we do not get rid of this evil, no medicine will ever save China.[6]

The trouble was with the Chinese people, not with science.

Mainland scholars are right: When he abandoned medicine for literature, "Lu Xun did not at all leave the natural science front. Throughout his life, natural science was always one of his combat weapons. His love for natural science never cooled. He always payed close attention to the development of natural science in our country, and he intuitively linked natural science to our country's fate."[7]

But that does not mean that he used the weapon of science scientifically. It certainly does not mean that he used the theory of evolution scientifically. We have already seen evidence of that in his theory of progress, his theory of human nature, his cry against "people eating," and his call for "dog beating." There is more evidence in a host of other themes.

Indeed there is evidence, already obvious in the above quotations, in the most common theme of all, the veritable theme song of all Chinese "Social Darwinists": China must struggle to survive. Even the variant, "the Chinese people must struggle to survive," was un-Darwinian.

Who must struggle? What was struggle? What was survival—or extinction? Lu Xun echoed many times over the "Darwinian" language of Yan Fu, Liang Qichao and so many other Chinese, whose greatest fear was that "the country perish and the race be extinguished (*guo wang zhong mie*)."[8] But for different reasons, neither of those dreadful possibilities was a true *Darwinian* danger.

"China" does not exist—except as an abstraction in men's and women's minds, and as "territory" delineated on maps and staked out, roughly, on the face of our globe. Physically, China includes, as the Chinese say, mountains and rivers (*shan he*), but it is mostly rock and dirt. Some of that rock and dirt will be washed out to sea, but until we blow up the planet, it will not perish from the earth, not even if "China" is "wiped off the map."

As a "state," "China," like all "states," has no biological existence, and is subject to no biological laws. However useful Darwinian idiom is in describing the "rise and fall" or the "birth and death" of states, it is useful only as analogy. Darwin did not study states, he discovered nothing about states. He proved nothing about states. States do not live and die. States are empty labels pinned in people's minds on groups of people ("groups" too are abstractions) or on the "governments" that rule such groups of people. In either case a "state" may "perish" without its people perishing. Therefore there is no necessary connection between the disappearance of a state and the extinction of a genetic group of people.

The survival of states was "immaterial" to Darwin. All that counted was the survival of people, or of races or varieties of people. So the only Darwinian question faced by the Chinese people was whether or not the Chinese people faced extinction—and the only Darwinian answer, in Lu Xun's day, was no.

Granted, all peoples face the *possibility* of extinction, the way the dinosaurs did. Our world, perhaps, is far more dangerous than theirs, for we face not only cosmic cataclysm but fire and brimstone of our own making—and a deluge of our own pollution. But even so, in Lu Xun's day, the Chinese people should have been the last people on earth to fear extinction, because they were then, as they are now, the fittest people on earth—as far as people can tell.

What Chinese faced was the threat of foreign conquest. But that was not the same thing as extinction. The Chinese had already been conquered for two hundred years, when they began to fear they might be conquered anew. But their Manchu conquerors never threatened their existence. Nor did they ever visibly effect the Chinese gene pool. It was the Manchus who faced the danger of assimilation (and even that was not extinction). The British might have conquered China. Other nations might have "carved up the Chinese melon." But the Indians survived in British India, and were fruitful and multiplied, as have the Chinese in Hong Kong and Macao. So conquest does not necessarily have anything to do with survival or extinction.

In the last year of his life, 1936, the year before the Japanese set out in earnest to conquer the Chinese, Lu Xun, rallying, so he hoped, all Chinese writers to the cause of resisting the Japanese, wrote that "In China the greatest question, the question for everyone, is the question of the survival of the race."[9] But the survival of the Chinese race was *not* in question. The Japanese proved, indeed, disgusting killers and rapists, but they would not, they could not, have carried out genocide in China. There were too many Chinese. There *was* Darwinian strength in Chinese numbers. The Chinese, enslaved, would still have been fit.

So Lu Xun's fears were not Darwinian.

Sometimes Lu Xun admitted as much. He once said, "I think the name "Chinese" will never disappear; as long as the race exists there will always be Chinese." His "great fear" was that the Chinese would have no "position" in the world.[10] Most clearly, he said elsewhere that "Chinese would still be Chinese" even if they were "slaves who lost their country." For "the existence of a state lies in its political sovereignty,"[11] but the existence of a people does not.

More often, however, he echoed the popular hyperbole:

Hereafter we truly have only two roads: One is to embrace our classical language and die, the other is to abandon it and survive.[12]

Our first duty is indeed to preserve ourselves. We only need ask whether things have the power to preserve us, not whether they are of the national quintessence.[13]

Although the world is not small, a vacillating race in the end will find no place in it.[14]

The word "extinction" can intimidate people, but it cannot intimidate Nature. For Nature is utterly unfeeling. If it sees a people bent on walking the road to extinction, it bids them proceed. It is not polite.[15]

If we look at things from the point of view of the human race, then if China reforms, it will be proof of human progress (because even a country such as this can reform). If it does not, it will still prove that the human race is advancing, because it is precisely because the human race is progressing that such a people cannot survive.[16]

Lu Xun spoke unscientifically about extinction and survival, but it was perfectly clear what he really feared and what he really wanted. He feared subjugation and he wanted freedom. Like all patriots he wanted his people free from foreign domination. That is "perfectly understandable." So why worry about a bit of hyperbole?

Why? Because Lu Xun's pseudo-scientific language has masked for many the totally idealistic nature of his thought, and thus has helped support a pseudo-scientific superstition.

The theory of evolution, as we have seen, cannot explain freedom, either as a reality or as an ideal. Lu Xun's dream of national freedom is meaningless unless free will exists, and evolution cannot explain the existence of free will (nothing can). Lu Xun's ideological struggle, therefore, not for existence but for an ideal existence makes him, once more, an idealist.

It may not "help" very much to label him an idealist, but it *will* help, if we can make it crystal clear that he was not at all a successful materialist.

Our final evidence of this lies in two final examples of Lu Xun's unscientific use of Darwinian idiom—first in his famous attempt to make monkeys of conservatives—or reactionaries:

> I think the theory that men and apes have a common origin is probably true beyond all doubt. But I do not understand why ancient monkeys did not all strive to become men, why instead they left descendants who to this day perform tricks for the amusement of men. Was there not one monkey in those days who wanted to stand up and learn how to speak in human speech? Or were there a few, who were attacked and bitten to death by monkey society for trying to institute change? Is that why they were never able to evolve?[17]

Lu Xun so liked that passage, published in 1919, that he repeated it, in expanded form, eight years later in 1927, in his famous speech to the cadets of the Whampoa Military Academy:

> Biologists tell us, "Men and monkeys are in no way very different. Men and monkeys are cousins." But why did men become men, and monkeys remain monkeys? This was simply because monkeys were unwilling to change—they *liked* to walk on all fours. Perhaps there was once a monkey who stood up and tried to walk on two feet, but the other monkeys must have said, "Our ancestors have always crawled. You are not permitted to stand." And they must have bitten him to death. They were not only unwilling to stand up, they were unwilling to learn how to speak—because they were conservative. But men were different. They finally did stand up, and spoke, and so finally were victorious.[18]

We need not belabor the ludicrousness of this parable as a description of evolution. Political satirists do not have to be scientific. This is witty, bitter, political satire. But even when translated into straight political language, Lu Xun's message was in no wise scientific. Lu Xun was expressing, as always, moral outrage. China's conservatives were killing China's progress. China's conservatives were literally killing China's reformers. They were benighted. They were inhumane. They were suppressing the enlightened, those who wanted to "speak like human beings," those propelled towards true humanity by the wheel of righteous dissatisfaction:

> Dissatisfaction is a wheel that rolls upwards. It can bear the unselfsatisfied human race forward towards humanity.
>
> Races with many dissatisfied people shall forever progress and forever have hope.
>
> Races with many people who only blame others and cannot fault themselves—face disaster.[19]

Was *dissatisfaction* the secret of progressive evolution? Did dissatisfaction have anything to do with the origin of species? Of course not. Mutation never came from dissatisfied DNA. Beasts did not engage in self-criticism. Races, for that matter, could not know which way was "up," unless "blessed" with Huxleyian intuition. Lu Xun's dissatisfaction came from pure Huxleyian idealism.

And so did his call for self-sacrifice, in his ultimate confusion of evolution and progress:

> I think the continuation of the race—that is, the continuation of life—is indeed a great part of the work of the biological world. Why does it want to continue it? Obviously because it wants to evolve and progress. But on the road to progressive evolution the new must always replace the old. Therefore the new should joyfully go forward, to grow up, and the old should also joyfully go forward, to die. If everyone thus moves on, that is the road to evolution.
>
> Old people, get out of the way. Urging, encouraging, let the young go. If there are deep abysses in the road, then use death to fill them up. Let the young go.
>
> The young will thank them for filling the deep abysses, so that they themselves can go on; and the old will thank the young for going on across the abysses that they have filled. —On, forever on.
>
> If one understands this, then from youth to maturity to old age to death, all should pass joyfully, step by step; for most shall be new people surpassing their ancestors.
>
> This is the wide, correct road of the biological world! The ancestors of the human race have all trod it before us.[20]

Again, as *science* this is absolute nonsense, as to a large degree we have already seen. The tenacity of life in this world is, granted, an amazing mystery, but unless there is a God, "the continuation of life" cannot be "the work" of the biological world. Unless there is more to this than meets the eye, "nature" is "up to" nothing. It does not *will* life to continue, or to evolve, or to "progress." There is no road to evolution. No species has ever had any idea where it was going. No species has ever said, joyfully or otherwise, to another species, "This way, please. Over my dead body." We may, of course, have been biologically programmed to sacrifice ourselves for our genes, but if so, as we have said, we are being pushed around. The continued life of our genes will do *us* no good. So why should we "joyfully go forward, to die?" If some *do,* their joy is a device. The young and the old owe each other no thanks—scientifically.

But both *can* feel thanks, or they think they can. As a political plea, or a moral plea, Lu Xun's cry does make sense, at least common sense, as a common cry echoed down the ages: "Elders, step down. Conservatives step aside. Give us a chance. Let things change. Let us try to make things better." That cry has seemed to many to make very good sense indeed in China, in 1898, in 1919, in 1989.

Politically, of course, it has not been unassailable. In the untitled little red book of Lu Xun's quotations published during the Cultural Revolution, not a single sentence from the above quotation was included—because the whole thing was "reformist." The Chairman did not believe that "the old" would "get out of the way." They would not "take themselves off the stage of history," for "anything reactionary," he said, "will not fall unless you hit it."[21] (So, ironically, the Chairman's followers who suppressed in 1989 the students who echoed Lu Xun's reformist cry, have, by their own logic, only set themselves up to be knocked down.) But here we need not argue about the political wisdom of Lu Xun's argument. He could support in any case, as we have seen, revolution or reform equally well with unDarwinian Darwinian argument ("That it has been possible to move from the first worm to mankind and from barbarism to civilization is simply because there has not been a single moment without revolution."[22]) What counts, once again, is that Lu Xun's exhortation to self-sacrifice was a *moral* argument. It was not based on any scientific law. It was based on a sense of good.

We need not here ask once again whether his sense of good made good sense, or, indeed, whether his moral sense made moral sense. Revolution and reform were two means to one end, a good end by definition: a good society, a "humane society."

But if his cause was just, did he hurt that cause by defending it with arguments that were not? Did he hurt his cause by defending it with pseudoscience?

I think he did, even though I would be the first to admit that there is some injustice in labeling him a pseudoscientist. Of course it is somewhat unfair to judge him by today's standards (although it is silly to *refuse* to judge him by today's standards). Of course much more is known about evolution today than in his day. And if even by the standards of his day he nowhere revealed any sophisticated knowledge of the workings of evolution, he may have known more than he revealed. For however great his respect for science, he was writing not for science but for his cause, and satire and sarcasm were his weapons. Therefore, in everything he wrote, we should grant him considerable poetic license.

Nonetheless, wittingly or unwittingly, Lu Xun did present much as science that was not science, and in so doing he unwittingly helped create or support, in the minds of undiscriminating readers, a superstition—that can be seen, at least in embryo, in passages in his own writings for which we *cannot* grant him poetic license.

This superstition was that moral positions can be equated with scientific truth.

In his early writings, Lu Xun turned first to natural science to find laws for human action. And so he used evolution, as he understood it, both to outlaw "people eating" and justify "dog beating." In his later writings he progressed, so he said, to social science. He began to talk of "social science" in 1930, first as an alternative to empty, unscientific talk of revolution, which had brought China, he said, nothing but grief:

> After that bitter lesson, to turn to seek a cure from a basic, true social science was naturally a definite advance.[23]

In another instance, in the preface to a book his brother Zhou Jianren had edited, entitled *Evolution and Devolution*, he offered "social science" as a cure to the problem of desertification in northwest China. Natural science, he said, could only explain the physical process. It could not get to the root cause of the problem which lay in social injustice:

> This book does not discuss that problem because it is limited to the realm of natural science. To provide a solution by going a step beyond that discussion of natural science there is social science.[24]

Scholars in the People's Republic have long maintained, with fair reason, I think, that "the 'social science' that Lu Xun spoke of was the science of Marxism,"[25] but whether it was or not, Lu Xun's discussion of social science did help people believe that Marxism was a science, because Lu Xun helped convince people that social science was as scientific as natural science.

But it is not. It never will be. It cannot be—unless, again, "existence," our existence, "is but an illusion,"[26] (and even then who are we to be deluded?). Social science is self-destructive. It cannot objectify man without objectifying the social scientist. But social scientists cannot objectify themselves, try as they might. For there is always a seeing self that separates itself from the seen self, or from the self that sees the self seeing the self. Mirrored in a mirror we cannot escape ourselves, but seeing, in that bondage we see our freedom. We cannot know ourselves unless we are free to know ourselves, but we cannot know ourselves unless the selves that we know are not free. So we cannot know ourselves, but we are free. In that way too we are "condemned to be free."[27]

I do not suggest that social scientists should therefore quit, only that they should remind themselves, periodically, of Zhu Xi's caution about *yi xin qiu xin* (using the mind to seek the mind).[28] In the natural sciences mind studies matter. In the social sciences mind, ultimately, studies mind. But who or what is it that *uses* mind to study mind? Mind cannot exist both as subject and object, if it is absolutely subject to natural law. So if we can study the laws of society, society cannot be governed by laws, absolutely.

True social science cannot exist if we do. True natural science cannot exist unless we do. So we must believe in our selves (even if we choose not to), and not too much in our social science.

But Marxists do believe too much in their social science. Marx believed he was a scientist. He believed he was *the* social scientist. And his followers have blindly kept the faith, a tragic faith because it has led over and over again to a ruthless self-righteousness that has struck at the very heart of Marxism. For Marx was not a scientist. He was a moralist, whatever that is. Like Lu Xun he beheld his world and saw that it was bad. He looked at his society and saw that it ate people. And he swore to change it. Everything he wrote thereafter, therefore, was based on a humanitarian *cri de coeur*, not on any scientific law. But fooling himself into thinking that his theory of history was a scientific law, and hardening his heart in the face of evil, he fooled others into thinking that they had a scientific right and duty (a totally unscientific concept) to annihilate "class enemies" and "counter-revolutionaries." And so Marx transformed his *cri de coeur* against people-eating (alienation) into a sanction for people-eating (class struggle).

In the name of Science, Marxists have eaten people as religiously as others have in the name of religion.

There are many roads to ruthlessness—the religious, the vindictive, the practical. The last could have sufficed, without any help from science. Lu Xun succinctly offered this practical plan for young people, as their "most pressing obligation":[29] "First, survival. Second, sustenance. Third, development. Anyone who dares hinder these three things we shall resist and exterminate," or, as he said elsewhere, "trample, under foot."[30] In the struggle for survival, ruthlessness was a "right" a million years before Marx and Darwin. But it was more easily justified than ever once Marx and Darwin made it part of a "scientific law," especially in China in the early twentieth century, where "Science" had a holy and unholy power for Chinese who had lost the faith of their fathers and felt forced to seek a faith from their enemies in the West. Darwinism rose above East and West. Marxism, though Western, was anti-Western. Natural science and social science both said all must struggle to survive.

So backed by Social Darwinism, Marxism, "for the people,"[31] ate people, in the name of Science, during Land Reform, the Suppression of Counter-Revolutionaries, the Anti-Rightist Movement, the Cultural Revolution, the Suppression of the Student Protest of 1989.

"Marxism is a scientific truth,"[32] said Mao Zedong. By nature, therefore, "no reactionary things will fall unless you hit them."[33] So "in suppressing counter-revolutionaries, please take care to hit them resolutely, accurately, ruthlessly."[34] For "our enemies are so fierce and cruel," said Deng Xiaoping, on June 9, 1989, "we should show them not one percent of forgiveness."[35]

That echoed, alas, Lu Xun's famous "last words" to his enemies: "I shall not forgive a single one,"[36] words recalled by Chen Boda, still, then, the Chairman's left-hand man, in October 1966, early in the Cultural Revolution, in his closing sermonette at the commemoration of the thirtieth anniversary of Lu Xun's death: "I think this is an important testament Lu Xun has given us, a testament that we must never let ourselves forget."[37]

Marxism was an unforgiving science.

In 1934, two years before his death, Lu Xun wrote a "random thought" lamenting that "science," supposed to save the nation, was being used in China, by conservatives, to defend the (people-eating) Chinese tradition. He quoted, in his lament, Mme. Roland, "'Liberty, Liberty, how many crimes are committed in thy name!,'"[38] fully intending that his readers substitute "Science" for "Liberty." He did not realize how many crimes would be committed later in Science's name, or how his own words would—and could—be used to justify them.

Lu Xun in the People's Republic

In the People's Republic, Lu Xun's words have been used to justify people-eating on a scale that Lu Xun in his darkest, loneliest, most pessimistic moments could never have imagined. For in death Lu Xun has supported the Chinese Communist party in everything it has done, transformed in death into what he never was in life: "a pillar of the establishment."

"The establishment" has changed. "The authorities" have changed, but always "the authorities" have claimed the backing of Lu Xun, ever since Mao Zedong canonized him in 1937 as "the Sage of New China," a convert-sage whose "thoughts, deeds, and works were all Marxismized [sic]" even though "he was not a member of the Communist Party organization."[39]

Thereafter, everywhere the Party went Lu Xun was sure to go, supporting the authorities, of the moment, and damning their real or imagined enemies.

But the authorities' Lu Xun has not been the only Lu Xun in the People's Republic. There has also been the scholars' Lu Xun, and the dissidents' Lu Xun, and the people's Lu Xun, all of which Lu Xuns have been many Lu Xuns.

Of these the simplest, and most simpleminded, has been the authorities' Lu Xun, as seen in the writings and speeches of high officials. Their Lu Xun has simply been for Mao Zedong, Marxism, the Communist Party, revolution, and class struggle, and against feudalism, imperialism, class enemies, and (the same thing) counterrevolutionaries, that orthodox line having been established by endless recourse to two handfuls of very carefully selected quotations.

The first handful, designed to prove that Lu Xun "ardently loved Chairman Mao and ardently loved the Communist Party,"[40] might seem to outsiders a bit skimpy. Granted, Guomindang censors did not make it easy for ardent lovers of Chairman Mao and the Communist Party to express their ardor, but Lu Xun mentioned Mao Zedong in *none* of his short stories or essays, and in only *one* of his seven hundred and ninety-eight letters. And in that he merely congratulated "Mssrs. Mao Zedong" for resisting the Japanese. It was in that context alone that he said, "I consider it an honour to call them comrades."[41] True, in a telegram to the members of the Chinese Communist Party's Central Committee, congratulating them for their successful, if costly, Long March, "Lu Xun and others" said, "On you rests the hope of China and humanity." And on some unspecified occasion Lu Xun reportedly said to a Japanese: "The Communist Party is the locomotive."[42] But that was the extent of his ardor.

The second handful of quotations is much fuller, for it contains all of Lu Xun's dog-beating injunctions. Chairman Mao used one in his very first pronouncement on Lu Xun, in his 1937 speech on the first anniversary of Lu Xun's death, in which he commended Lu Xun for showing "not a bit of the fake pity of a hypocritical *junzi*."[43] But the Chairman was only calling, at that point, for no pity to be shown to the Japanese. Chen Boda, as we have seen, sang a similar song on the thirtieth anniversary of Lu Xun's death, but he called for no forgiveness to be shown the Chairman's enemies in the Cultural Revolution, those former authorities turned dogs, Liu Shaoqi, Deng Xiaoping, and the rest. That the "Gang of Four," and the Chairman, should call Chen Boda himself a dog, however, four years later, and that the twice-beaten Deng Xiaoping should crawl a second time out of the water to call the Gang dogs, six years after that, should prove how useful and how dangerous Lu Xun's injunctions could be. Anyone could quote Lu Xun against anyone.

"Who are our enemies? Who are our friends?"[44] asked the Chairman. He did not ask, "Who is to say?"

Dissidents have used Lu Xun differently, to back them up in their criticisms of the authorities. Although the authorities have done their best to monopolize Lu Xun, they have clearly failed. For Lu Xun has been the patron saint of dissidents all along, ever since the first great conflict between the party and disgruntled intellectuals (after, that is, the party's conflict with Lu Xun himself) in 1941–1942. We cannot trace here the history of "dissent" in the People's Republic (or struggle with the definition of the word "dissident"[45]), but three well-known cases can establish the continuity of Lu Xun's influence.

In 1941–1942, in Yan'an, leftist writers of good standing, Ding Ling, Xiao Jun, Luo Feng, Wang Shiwei, and others, raised Lu Xun's banner in an ill-fated attempt to assert (or reassert) the (traditional, Confucian) intellectuals' moral right, and responsibility, to criticize error wherever they saw it—even in the policies and actions of their leaders. To do that, said Ding Ling, "We need the *zawen*," the "random essay" that Lu Xun had forged into a biting weapon of social or political criticism. "We have abandoned our responsibility," she said: "Lu Xun is dead, and all of us often say we should do this or do that in his memory. But we lack the courage to follow him in acting without fear of trouble. Today, I think the best thing we can do is to follow his example in steadfastly, constantly, facing the truth, and in daring to speak for the truth, fearing nothing. Our age still needs the *zawen*. We should not give up that weapon."[46]

But the Chairman, in his infamous "Talks at the Yan'an Forum on Art and Literature," outlawed *zawen*, unless aimed at "enemies of the people"—as defined by him. Ding Ling and her colleagues were denounced— and then forced to denounce themselves and each other. And then they were given their first taste of "reform through labor." Wang Shiwei, who refused to denounce himself, or others, was finally shot—or beheaded.

After that, after that first "Rectification" campaign, and then "Thought Reform," and then the mind-numbing "Anti-Rightist Movement" of 1957, it took courage indeed, or foolhardiness, for Wu Han to resurrect the *zawen* in 1959. But he did—in double tribute to Lu Xun.

He called his essays *zawen,* and even more pointedly "Spears," which is what Lu Xun said short essays should be.[47] Lu Xun said, "If we have spears, then let us use spears. We do not have to wait for tanks or incendiary bombs, which are now being built or which will be built."[48] Wu Han daringly took up Lu Xun's metaphor and indeed his point, that writers in fulfilling their (Confucian) responsibility, should use whatever "weapons" were at hand: "If you have a spear you have to use it. What can you do? You can't put it down. Why not throw it? What does it matter? If it hits the target, it may at least cause some discomfort."[49]

And so he carried out a campaign of veiled criticism—for which he paid with his life.[50]

Wang Ruowang was more fortunate. A "dissident" writer from his days in Yan'an, where he worked with Ding Ling and Wang Shiwei, he wrote *zawen* when the Hundred Flowers bloomed—and was then attacked as a "rightist." During the Cultural Revolution, he was imprisoned as a "counter-revolutionary." But he survived—to be "rehabilitated" in 1979 and expelled from the Communist Party in 1987, for writing another batch of *zawen* (and works of fiction) an amazing batch, almost as much as all he had written before he was silenced in 1962.[51]

In all of this Wang Ruowang was a follower of Lu Xun. The "wall paper" that he edited in Yan'an took its name from another of Lu Xun's metaphors for *zawen,* "Qingqidui," the light cavalry.[52] One of the *zawen* he wrote in 1980, "The Relationship between Literature and Politics is not one of Subordination," echoed the old argument of Wang Shiwei and of Lu Xun before him.[53] Wang Shiwei wrote his *zawen,* "Politicians-Artists" in 1942.[54] Lu Xun gave a speech on "The Divergent Roads of Literature and Politics" in 1927.[55] The dissidents' Lu Xun has always said, first and foremost, that writers *must* be dissidents.

It would take a separate volume, to pursue the dissidents' Lu Xun in detail, but at least we can see that in a fundamental sense, which we must seek to identify in a minute, the dissidents' Lu Xun has always rejected the radicals' dictum (he did so before they coined it): *zhengzhi gua shuai* (politics takes command).[56]

It would take another volume, an even thicker volume, to pursue the people's Lu Xun, to learn what Lu Xun has meant to "the masses," to the literate masses at least (to the illiterate masses he cannot have meant much) to "intellectuals," students, schoolchildren and any workers and peasants who read, and who might have read Lu Xun's stories or essays or essays about Lu Xun. But we cannot search for the people's Lu Xun until the authorities let us, until they warm to the technique of social-scientific surveys of public opinion, and encourage members of the public to make known their opinion. So we must wait for new authorities.

But even without being socially scientific we can safely divide those who have thought at least a little of Lu Xun into three groups (without knowing the relative size of those groups). Some must always have accepted the authorities' Lu Xun: the Chairman's Lu Xun, the Party's Lu Xun, the Gang of Four's Lu Xun, the "Reformers'" Lu Xun. Others must have applauded quietly the dissidents' Lu Xun. But if we can believe a common lament of Lu Xun scholar's, after the Ten Years of Disaster a third group appeared, that rejected the authorities' Lu Xun *without* accepting the dissidents' Lu Xun, a group of "young people," who were "sick and tired" of Lu Xun.[57]

I have met such people. They exist, or did exist. No one knows in what numbers, but I believe in "significant numbers," because officials, scholars, and scholar-dissidents all complained of their existence throughout the entire great decade of reform, from 1979 to 1988. Some seemed to fear that a whole generation might have lost interest in Lu Xun—or worse. "Many young people who grew up in the decade of disaster,"[58] treat Lu Xun with "cold indifference," said one leading Lu Xun scholar, Lin Fei (Pu Liangpei): "Some even feel a certain antipathy for Lu Xun's works, although they themselves cannot say why."[59]

But to others it was perfectly obvious why. Their apathy and antipathy was all the fault of "the Gang of Four." In the decade of disaster, said Zhou Yang, the Gang of Four "ruined"[60] Lu Xun, by "stealing his banner," said Han Shaohua, "to cover up their evil deeds."[61] Completely distorting Lu Xun by quoting his works out of context,[62] the Four "forced Lu Xun to go out and serve them in their sinister schemes,"[63] to "seize control of the party."[64] They turned Lu Xun "into a god or demigod," and "Lu Xun's works became their 'bible.'" Indeed, said Zhu Zheng, "they seemed to be able to find in Lu Xun's works a basis for their every outrage, for their every evil or stupid act."[65]

Cynically they sought to "deify" Lu Xun. Actually they "uglified" him.[66] They "murdered [him] with praise."[67] For what they really did was use Lu Xun's works "as tools," to "beat people," and "rectify people."[68] "They used Lu Xun as a bludgeon," said Yi Zhuxian, "to beat all the people they wanted to knock down,"[69] "to attack," that is, said Zhou Yang somewhat self-servingly, "all who opposed them or disapproved of them," all "true revolutionaries."[70]

Let us ignore that most "metaphysical" of Marxist squabbles: who were the true revolutionaries? In all other respects, members of "the Gang" (for whom we should read those in charge, as much as any were in charge, of the Cultural Revolution) were guilty as charged: The Chairman and his party made cynical use of Lu Xun in a whole series of campaigns against the party in charge of the party, the party of Liu Shaoqi and Deng Xiaoping.

The Chairman himself urged people to "read a little Lu Xun"[71] (*Nota bene,* not too much). The Gang made up a reading list—of weapons, chief among which, it will come as no surprise, was Lu Xun's essay "On the Advisability of Postponing Fair Play." That essay, recalled Han Shaohua, "was used as an official circular, reprinted in papers throughout the country."[72] It was quoted ad nauseam in the little red book of quotations from Lu Xun. It was the very first selection in a *neibu* (for internal distribution only) selection of Lu Xun's *zawen*.[73]

Through Lu Xun's essay "On the Advisability of Postponing Fair Play," the Gang did, indeed, urge people to beat people.

Consider this typical piece of Lu Xun "scholarship" from the early seventies, when the Gang was struggling to revive the Cultural Revolution after the Chairman had been forced to send in the army to rein in the Red Guards and restore order. In answer to his own question, "Why recommend the study of some of Lu Xun's *zawen?*," one Lei Jun replied:

> Today, the swindlers, Liu Shaoqi and his ilk, have already been beaten into the water by the revolutionary populace, but their poisonous influence has not yet been wiped out. To uphold Lu Xun's revolutionary spirit of "beating dogs in the water," and thoroughly denounce the counter-revolutionary revisionist line and the staggering crimes of the swindlers, Liu Shaoqi and his ilk, is one of the urgent combat missions now before us.[74]

After that, as Han Shaohua, Yi Zhuxian, and many others have pointed out, the Gang enlisted Lu Xun's help in the (thinly) veiled criticism campaigns against Deng Xiaoping and Zhou Enlai, the "Criticize Lin Biao and Confucius" campaign, the "Evaluate the Legalists and Criticize the Confucianists" campaign,[75] the "Criticize *The Water Margin* and Oppose a Restoration" campaign, the "Criticize the Bourgeois Democrats" campaign, and others.[76] Proof lies in such volumes as *Lu Xun pi Kong fan Ru wenji* (Selected Lu Xun texts criticizing Confucius and opposing Confucianists) and *Jicheng Lu Xun de fan Kong douzheng chuantong* (Carry on the tradition of Lu Xun's struggle against Confucius).[77]

During "The Decade of Disaster," Chairman Mao and "the Gang of Four" used Lu Xun's works to help justify their attacks on most of China's highest party leaders. They also used "Lu Xun's revolutionary spirit of 'beating dogs in the water'" to justify the mental or physical beating of millions of other people.

That explains Han Shaohua's lament, written in 1982:

I remember a few years ago, when I spoke with some young people in their early twenties about Lu Xun, they either said not a word (I do not know whether they had nothing to say or whether they had other reservations), or else they said such things as "Lu Xun was pretty brazen. He was in every political campaign," or "Old Lu Xun beat every one he could get his hands on. With his pen he was positively ruthless," or "Otherwise, why would he have shouted 'Great men are cruel.'" —But what about "True heros are not necessarily unfeeling?" Was not that the compassionate feeling that welled up from the depths of Lu Xun's heart to pour out through his pen? —This really is dreadfully disheartening. Truly it leaves one with nothing to say. There is no denying it. Lu Xun had indeed fallen into a tragic state, unimaginable in his life time.[78]

And Lu Xun research, other Lu Xun scholars complained, had fallen into a similarly depressed and depressing state. In 1980 some insisted that "because the 'Gang of Four' seriously distorted and exploited Lu Xun, hurt his image, and dampened young people's enthusiasm for studying his works, Lu Xun research, it seems, is no longer popular either, and certain publishing houses are reluctant to publish studies of Lu Xun."[79]

Five years later one could hear the same complaint:

In society at large are there not already people who have made it clear that they are tired of Lu Xun scholarship? Are not some journals unwilling to publish any more academic articles on Lu Xun, and some publishing houses unwilling to publish any more academic books on Lu Xun? Today, is not the publishing of an article on Lu Xun, or a book on Lu Xun, virtually seen as a stupidly impractical "act of philanthropy?"[80]

But there is something odd about these complaints. They were made in published books that were part of a decade-long explosion of Lu Xun research, during which tons of books and articles on Lu Xun were published.

From 1949 to 1966 "almost a hundred"[81] books were published about Lu Xun. Then, during the Cultural Revolution, in what Zhang Mengyang calls the "counter current" period of Lu Xun research, "normal scholarly research almost came to a complete halt,"[82] except that it was still possible, given "The Gang's" rabid championing of Lu Xun, for scholars in a position to do any work at all to work on research materials. In that field, great work was done, as Lin Fei noted, caustically, in 1981:

> Under the cultural despotism of 'the Gang of Four,' if one published a scholarly critique in any way not to their taste, one was immediately censured. Work on research materials was less likely to be attacked. Also at the time, in the field of modern literature, the vast majority of authors were still "off limits." Lu Xun was the only one one might be permitted to discuss, so the fruits of work on Lu Xun research materials appeared in great quantity in quite a few scholarly journals. This was somewhat like the brutal literary inquisition of the Kangxi and Yangzheng periods in the Qing dynasty—people rarely wrote critical studies, which could easily spell disaster, but the period did inspire a flourishing of empirical research.[83]

But the fall of the Gang of Four led to a "renaissance"[84] of critical Lu Xun studies. Indeed the Academy of Social Sciences initiated an unprecedented national research effort. Research units were organized in every province in the land, save Taiwan, and hundreds of scholars chose to, or were chosen to, or assigned to, study Lu Xun. And they did.

Conferences were held all over the country, and preparations were made for a superconference to celebrate, in September 1981, the one hundredth anniversary of Lu Xun's birth.

In the year before that conference, the year in which "some scholars" complained that certain publishing houses were reluctant to publish works about Lu Xun, at least seventeen monographs were published and over three hundred scholarly papers. In the year of the conference itself, Lu Xun scholars, and publishing houses, went wild. In addition to a new edition of the (ever more) *Complete Works of Lu Xun*, at least one hundred and nineteen major books were published on Lu Xun and over three thousand scholarly articles (not to mention thousands of newspaper articles),[85] more than all those published in the twenty years before 1949 and over half as many as all those published in between 1949 and the Cultural Revolution. 1981 was, needless to say, a "big year" for Lu Xun research, not likely to be repeated.[86] But research continued. In 1984 Zhang Mengyang, although lamenting the relative scarcity of studies of Lu Xun's *zawen*, reported a veritable "sea" of works on Lu Xun in general, "absolutely uncountable."[87] And two years later, in 1986, the fiftieth anniversary of Lu Xun's death there was another "big year."

Scholarship on that scale should have proved depressing only to foreign would-be scholars of Lu Xun, who felt they somehow had to confront it all.

But did all that scholarship prove that reports of Lu Xun's unpopularity were exaggerated? Not necessarily. Publishers might well have thought that too much had been published about Lu Xun—especially if all those volumes failed to sell, or sold only to Lu Xun scholars—not to young people. What did it mean, if Lu Xun was unpopular despite all that research?

In 1988, one of the most promising of the younger Lu Xun scholars, Wang Hui, in a most important, and controversial, article, entitled "Lu Xun yanjiu de lishi pipan" (An historical criticism of Lu Xun research), echoed the common complaint of seven years before—but with a threatening new twist.[88]

Despite the aforementioned "sea" of studies of Lu Xun, Wang Hui lamented that "people cannot help lamenting the cold and lonely state of 'Lu Xun research.'" But he did not blame that state on "the Gang of Four." He blamed it much more broadly on *"the leaders of China's political revolution"* (his italics) and on *China's Lu Xun scholars themselves* (mine).[89]

Wang Hui likened "Lu Studies" to "a solemn, 'ancient citadel' towering, thanks to its glorious history and its solid internal structure, over the confused, complex, and radically changing plain of modern thought and literature." But "some," he said, "have not unmockingly called 'Lu studies,' 'classical research,' because however conscientiously and seriously scholars have continued to add bricks and tiles to that 'ancient citadel,'" their research "has not established a dialogue with modern life."[90]

A devasting indictment. It was not just the Gang's Lu Xun who had alienated China's youth: *The scholars'* Lu Xun, he said, *had nothing to say to the present.*

Why not? Because the scholars' Lu Xun was still prisoner to politics—as he had been since the founding, since even before the founding, of the People's Republic.

Lu Xun research, said Wang Hui, had been born in the politics of the Communist revolution, and it had been subordinate to the politics of that revolution ever since:

> The most important historical tradition and the basic framework of Lu Xun research were formed in that period. At that time Lu Xun's image was established by the leaders of China's political revolution as one of that revolution's ideological or cultural authorities. Basically speaking, everything Lu Xun research has done since then has only perfected and enriched the image of that "new culture" authority. As a result, the demands of political authorities on the corresponding ideological authorities have become the highest conclusions of Lu Xun research, and Lu Xun research itself, whether or not the re-

searchers involved have realized it, has all been of a given political ideology.[91]

Wang Hui did not use the Gang's notorious slogan, *zhengshi gua shuai* (politics in command), but surely he thought that "politics in command" had been the bane of Lu Xun research since the beginning—and the cause of Lu Xun's present silence.

And in the beginning was the party and the Chairman not the Gang. Wang Hui decried, as so many others had, the party's "deification" of Lu Xun, but he did not use that term. Much more pointedly he lamented the "sanctification" (*shenghua*) of Lu Xun,[92] tracing that term to its source, the aforementioned speech in which Mao Zedong, on the first anniversary of Lu Xun's death, called Lu Xun a "first class *shengren* (sage)"[93]—a "sage" because he recognized the truth of Marxism-Leninism (and the Thought of Mao Zedong).

Anyone who reads the Chairman's speech can see what the Chairman meant. Lu Xun "was not a member of the Communist Party organization," said the Chairman, "but his thought, his action, his works, all were Marxismized." That explained "Lu Xun's value in China."[94] What did *that* mean? It meant that Lu Xun *had no other value*. It meant that all Lu Xun had to say of value was that Marxism was right. He had nothing else *of value* to say. That is why the "sanctification of Lu Xun" silenced Lu Xun, and why Lu Xun research, almost all of which helped support the sanctification of Lu Xun, helped silence Lu Xun.

In the first forty years of the People's Republic, Wang Hui was the first to say that outright. "Once Lu Xun accepted Marxism," he said, "he seemed to enter the realm of the sacred and the absolute"—and that was the end of Lu Xun research, "because there was absolutely nothing we could say to the sacred and the absolute." Thereafter, "all we could see was Lu Xun's 'identity' with us and our age—because we and our age were Marxist and socialist."[95]

In the course of his criticism Wang Hui made a bold, if classic (and classically bold and classically foolhardy) "dissident" appeal for "pluralism," for the blooming of a hundred flowers, although he avoided that oft-planted and oft-uprooted phrase. He denounced all unified systems of thought, all "unitary, all-explanatory theories":

We should point out that in the development of human thought all unified theories that explain everything, all normative ideological systems, are necessarily coercive. They necessarily rely on religious or political authority. Under such ideologies the independent individual can no longer think freely. He must and can only explain anything he

encounters according to such an ideology's prescribed norms. Therefore his conclusions are predetermined. They are proofs of the au-
thoritative ideology's preconceived concepts. Thought, that hawk
that can only soar in a free universe, loses its freedom in the cage of
unified concepts. What else can it do but silently sigh?[96]

Even in the encouraging days of 1988, that was a courageous paragraph in a courageous article, courageously published—as the leading
article—in *Wenxue pinglun* (Literary review), edited by Liu Zaifu, one of
the very best of the Lu Xun scholars half a generation ahead of Wang Hui.
When Wang Hui said that the kind of religious faith in a ruling ideology that he decried came out of the period of the Communist
revolution, a period "not far removed in years but so foreign to people living in the modern age, who take pluralism and independence as obvious
standards of value, that it seems a whole century away," he came dangerously close to publicly challenging Deng Xiaoping's "Four Insistences,"
his insistence that Chinese must "insist on" or "resolutely uphold" four
seemingly specific but actually very fuzzy "basic principles": The socialist
road, the dictatorship of the proletariat, the leadership of the Communist
party, and Marxism-Leninism and the Thought of Mao Zedong.[97]
Wang Hui's indictment of all ruling ideologies must surely have
seemed the most "important" part of his article to the orthodox, but more
important to us, at least for the moment, in our search for the real Lu Xun,
is his indictment of his fellow Lu Xun scholars. Wang Hui said that "Lu
Xun, that richly complex giant, rooted in Chinese history and
society, . . . lost in 'Lu studies' his ability to carry on a dialogue with the
whole series of problems of modern life."[98]
Exaggerated? Unfair? Of course, to a degree. But also, to a degree,
tragically true.
We should first say some things in defense of Wang Hui's
colleagues:
Surely it was not necessarily the fault of Lu Xun scholars that there
was little dialogue in the 1980s between Lu Xun and young Chinese. As
Lin Fei said, the sorry state of education during the Cultural Revolution
ill-equipped a whole generation to read Lu Xun.[99] Moreover, if the next
generation read neither Lu Xun nor books about Lu Xun it was perhaps
because the decade of relaxation allowed them at last to do what Lu Xun
had told Chinese students to do sixty years before: "Read fewer—or do
not read any—Chinese books, read more Western books."[100]
In the 1980s many Lu Xun scholars clearly *tried* to "open up a dialogue" between Lu Xun and the "problems of modern life." At the
conference celebrating the one-hundredth anniversary of Lu Xun's birth,

Lin Fei, addressing "the tasks we face," insisted that "Lu Xun research must be of practical service." He acknowledged that when some scholars heard that phrase, they became "a bit worried and nervous," remembering the use to which the Gang of Four had put those words. But clearly the practical service he imagined was of a very different sort from that of the Gang of Four.[101]

Over and over again, throughout the eighties—at least until June 1989—Lu Xun scholars tried to draw attention to the "practical significance" of Lu Xun's attacks on "feudalism" and "the national character." The common view—and the simple truth—was that reported by Wang Junji as the view of "some comrades" at a joint meeting in 1980 of the directors of the Lu Xun Research Association and the editorial committee of *Lu Xun yanjiu* (Lu Xun research):

The social ills that Lu Xun pointed out, in many corners, still exist today.[102]

I believe the corners "some comrades" had most prominently in mind were in high places. Lin Fei said that the Gang's "literary inquisition" proved that "the elimination of feudal vestiges is extraordinarily difficult." He also said it proved "how necessary it is to oppose feudal authoritarianism and strengthen socialist democracy."[103] In the decade following the fall of the Gang of Four, that refrain would rise in an evermounting chorus—until June 1989.

When Chinese students shouted for reform, they did not just shout for economic reform. When Lu Xun scholars said, "If we are to reform this country of ours, we must not cast aside Lu Xun's banner, or his thought,"[104] they were not just thinking of economic reform either. Why then should there have been "no dialogue" between Chinese students and the Chinese Lu Xun scholars' Lu Xun?

Were China's Lu Xun scholars too timid? Westerners who have never faced the very real dangers that Chinese scholars have always faced should be cautious in calling Chinese scholars timid.

Granted Lu Xun scholars have usually been indirect in their criticisms. They have been far better at hinting that Lu Xun's works were of practical significance than at spelling out that significance. But that caution was the better part of valor. They were careful perforce—as Lu Xun had been before them.

Chinese scholars, as a group, have lived between the officials and the "dissidents." Some have done their best to help create and propagate the official Lu Xun, meeting the demands of the moment. Others have cautiously, but courageously, leaned toward dissent, becoming "dissidents" themselves when caught "out of bounds," more often than not by official redrawings of "the boundaries"—in invisible ink.

Under such conditions, Lu Xun scholars in the eighties were not timid. Wang Hui knew that. He knew too that the best of them had done serious, sophisticated work throughout the eighties, in biography, bibliography, aesthetics *and* interpretation. That is why he apologized to his colleagues in the middle of his critique.

> I must apologize to those serious scholars. Under the set conditions of the times they have made a great number of contributions to Lu Xun research that still are of important value to this day.[105]

But even so he stuck to his indictment. And he was right. However much Chinese scholars may have *wanted* to let Lu Xun speak, their academic theories, "under the set conditions of the times," helped to gag him—especially their theories of his evolution.

The Myth of Lu Xun's Evolution

The myth of Lu Xun's evolution has dominated analyses of Lu Xun's thought ever since 1933, when Qu Qiubai, alas with Lu Xun's blessing, proclaimed that Lu Xun, cruelly enlightened by the Nationalist's purge of the Communists in 1927, "finally moved from the theory of evolution to the theory of class struggle." In 1927, Lu Xun "*progressed* from the theory of evolution to the theory of class struggle," and so, at last, "took to the road of true revolution."[106]

In the orthodox Marxist lingo repeated in a thousand and one re-tellings of Lu Xun's Paulinesque conversion, Lu Xun was said to have made a revolutionary "flying leap" *(fei yue)* at a traumatic moment in his intellectual evolution. After a slow evolutionary process of "quantitative change" *(liang bian)* he was transformed in a revolutionary moment of "qualitative change" *(zhi bian).* And so he turned, or was turned (a difficult Marxist problem) from an evolutionist, whatever that was, into a Marxist.

Qu Qiubai did not, of course, call Lu Xun a Marxist outright. Nor, writing under a pseudonym, did he confess that he himself was one (that he had been, indeed, the secretary of the Communist party), but he did not have to. Who else but the Communists believed that class struggle was what made the world go round?

Qu Qiubai's "Preface to a *Collection of Lu Xun's Random Thoughts*" was of immediate political significance, therefore, to non-Communists and Communists alike. For it offered the first unambiguous Communist support for Lu Xun and established the first unambiguous Communist claim to Lu Xun.

Before then, Communist writers had ridiculed and rejected Lu Xun, the Creation Society and Sun Society crowds (one assumes to their undying regret) having dismissed him in 1928 (a year *after* his alleged conversion, which clearly they failed to detect) as a "noxious feudal remnant," a "two faced counterrevolutionary,"[107] and an "unsuccessful fascist."[108]

True, Lu Xun's first close Communist friend-to-be, Feng Xuefeng[109] came to his defense in May 1928, seven months before they met. In an article entitled "Geming yu zhishi jieji" (Revolution and the intellectual class)—a Communist defense, of sorts, against a Communist attack, he tried out, for the first time, "scientific" Marxist theories of literature,[110] gleaned from Japanese translations of Marxist works that he was busy re-translating into Chinese. Having mastered Marxist smugness, at least, he condescendingly defended Lu Xun as, in effect, a junior varsity revolutionary, one of that "second class of intellectuals" that had "accepted revolution, and was drawn to revolution, and yet at the same time looked back to the old, and still loved the old." Intellectuals like Lu Xun were wont to be too passive, said Feng Xuefeng, but "the revolution will not be hindered by them."[111]

That defense was so unambiguously ambiguous that Lu Xun is said to have said, upon reading it: "This guy is basically on the side of the Creation Society!"[112]

Nonetheless with Rou Shi[113] defending in turn Feng Xuefeng, and introducing him to Lu Xun, Lu Xun and Feng Xuefeng soon became true friends, whereupon Feng Xuefeng tried to defend Lu Xun less ambiguously. His article "Fengci wenxue yu shehui gaige" (Satirical literature and social reform), published in May 1930, is now said to have prepared the way for Qu Qiubai's "Preface,"[114] for in it Feng Xuefeng insisted that satirical literature, of which even Lu Xun's enemies acknowledged Lu Xun to be a master, was potentially "one of the sharpest forms of the literature of class struggle."[115]

Still, it was Qu Qiubai's "Preface" that was hailed as the first true "milestone"[116] in Lu Xun studies, for it was the first study to "document" Lu Xun's conversion to the theory of class struggle.

Qu Qiubai's documentation came from three unfortunate sentences that Lu Xun wrote in 1932.

Five years to the month after Jiang Jieshi did his best to massacre the Communists, Lu Xun declared that that massacre had changed his philosophy:

I left Guangzhou in 1927 scared speechless by the blood.

I had always believed in the theory of evolution; I had always thought that the future must surpass the past, that the young must surpass the old. . . . But later I realized I was wrong. . . . When I was in Guangdong I saw with my own eyes the reality of young people dividing themselves into two great camps, some secretly reporting people, some helping officials arrest people. Because of that, my line of reasoning was blown to bits.

There is one thing for which I must thank the Creation Society: They 'pushed' me into reading some scientific theories of art and literature. . . . And consequently I translated Plekhanov's *The Theory of Art*, to correct my bias—and through me the bias of others—of only believing in the theory of evolution.[117]

Lu Xun meant, said Qu Qiubai, that he had seen in the Nationalist's "party purge" the bloody truth: One could not just wait for evolution, for the fitter young to inherit the earth from the less fit old. The way was revolution, and revolution was class struggle, not generational struggle. The Communists, that is—he did not have to say so outright—were right.

But was that really what Lu Xun meant? Probably.

When Qu Qiubai was writing his "Preface" he supposedly "saw Lu Xun almost every day and freely talked about everything."[118] Lu Xun, when he read Qu Qiubai's completed manuscript, was supposedly so pleased and excited that "he forgot about his cigarette until it burned his fingers." So said Qu Qiubai's wife, Yang Zhihua. She also said Lu Xun was "very much satisfied" with the Preface. Feng Xuefeng later said that Lu Xun said: "Its analysis is correct."[119]

Feng Xuefeng later reconstructed other quotations from Lu Xun that would be cited a thousand times to prove Qu Qiubai correct:

This time it was young people who taught me a lesson. —I believed in the theory of evolution. I thought young people would always surpass the old. I thought that those in the world who oppressed and killed young people were probably old people, but as the old people would die off first, the future would still be a bit better. But it's not like that. It's young people who kill young people, or inform on them, or themselves arrest them. In the past when warlords killed young people, I used to feel grief and indignation, but this time, before I could feel grief and indignation, I was scared out of my wits. My theory of evolution was completely bankrupt.

The theory of evolution helped me. It did point out a way. To understand natural selection, to believe in the struggle for existence, to believe in progress, was certainly better than not to understand, and not to believe. I just did not know that the human race had class struggle.

As I said, Darwin too was concerned with society. He just failed to recognize society's class opposition and class struggle.[120]

If we can trust Feng Xuefeng's memory, and honesty, and I do not know why we should not, Lu Xun did indeed seem to have accepted Qu Qiubai's analysis. And for forty years after Lu Xun's death, so, in Communist circles, did everybody else.

On October 19, 1937, the first anniversary of Lu Xun's death, Xu Guangping and Feng Xuefeng, sharing a speaker's platform somewhere in Shanghai, both echoed Qu Qiubai's thesis. Feng Xuefeng, indeed, echoed Xu Guangping's echo: "Xu Guangping said just now that Lu Xun originally believed in the theory of evolution, but later firmly believed in the theory of class revolution." Then Feng Xuefeng added an important element himself: "Fighting for the race and the masses with a noble love for the people, it was only natural that Lu Xun should believe in the theory of evolution, and *inevitable* that he should progress to the theory of class revolution."[121]

On the very same day, as we have seen, Mao Zedong canonized Lu Xun in far off Yan'an, hailing him as a man who "came out of feudal society in its collapse" to be "Marxismized."[122] And Lu Xun, said Mao Zedong three years later, "represents the vast majority of the race. —Lu Xun's direction is the direction of the new culture of the Chinese race."[123]

So Qu Qiubai's thesis and the Chairman's sense of direction went together to make Lu Xun's evolution both prophesy and proof of the inevitable evolution of the Chinese people—from feudal ignorance to Marxist enlightenment.

For forty years, most Lu Xun scholars accepted Qu Qiubai's thesis without question. Li Helin, for example, exulted in Qu Qiubai's insight without the slightest fear of contradiction:

In the last twenty years or more, over a thousand essays have discoursed upon Lu Xun, but not one can compete in beauty with Qu Qiubai's. In China's progressive literary circles, this is universally acknowledged.[124]

In the "renaissance" of Lu Xun studies after the fall of the Gang of Four, however, as if to test the new waters of "free discussion,"[125] Lu Xun scholars began to challenge Qu Qiubai. In the incredible flurry of scholarly activity that marked the hundredth anniversary of Lu Xun's birth, hundreds of articles—and many books—attacked Qu Qiubai's famous formula that "Lu Xun progressed from the theory of evolution to the theory of class struggle." The best scholars asked, as they surely should have, what Qu Qiubai—and Lu Xun—*meant* by "the theory of evolution." They questioned the notion that "the theory of evolution" could "characterize" the "early period," or former period" of Lu Xun's thought, or indeed the "middle period." They denied, however, that his belief in the theory of evolution was ever "blown to bits." At best they found Qu Qiubai's thesis too simple. They sought to plot out with greater precision the steps in his "progress" toward materialism and Marxism. They argued endlessly with those who still found Qu Qiubai's thesis "basically correct,"[126] and with each other, about the precise moment at which Lu Xun "completed" his "flying leap."[127]

I cannot here do justice to their arguments. They wrote so much I cannot here even summarize their arguments. I can only echo twice over Wang Hui's apology to the many "serious scholars," and good scholars (on both sides), who raised indeed the study of Lu Xun's thought to new levels of sophistication.

But I must criticize them all.

I would side with the critics of Qu Qiubai if his critics and defenders did not remain on the same side.

Wang Junji and Gao Mingluan said in their able summary of this whole debate:

> Lu Xun's thought went through a process of change from revolutionary democracy to Communism, from quantitative change to qualitative. This everyone admits.[128]

Then Wang Junji and Gao Mingluan listed Chinese grounds for disagreement: eight periodization schemes for Lu Xun's progress—or evolution.

But all eight schools were united in their assumption that periodization was *the* topic in the study of Lu Xun's thought, united in their assumption that *the* story was the *development* of Lu Xun's thought.

Without challenging the dogma that Lu Xun in his last decade was "pro-Communist," *without* challenging the Chairman's contention that at the last "his thought, his action, and his works were all Marxismized," I wish to argue that the development of Lu Xun's thought is of no great significance—because there was no significant development in Lu Xun's thought.

That is why I wish to argue that if we are to "liberate" Lu Xun, we must "blow to bits" Qu Qiubai's thesis, Lu Xun's self-analysis, and Chinese scholars' "bias of believing only in evolution." We must "blow to bits" the myth of Lu Xun's evolution.

It will seem presumptuous, or worse, to reject Lu Xun's self-analysis, and to reject Qu Qiubai's thesis, when Lu Xun accepted it. But although Lu Xun said, "I know myself; I dissect myself no less mercilessly than I dissect others,"[129] he was not his own best critic.

Lu Xun's self-analysis and Qu Qiubai's thesis both foundered on an irresponsible use of that wretched mistranslation that has plagued Chinese discussion, and our discussion, of "Lu Xun and Evolution" since the beginning, that wretched term, *jinhua lun* (literally the theory of progressive change), used indiscriminately to refer to Darwin's theory of *evolution* (understood or misunderstood) and to all sorts of theories of progress. As we have sadly noted before, no term in twenty-five hundred year's of Chinese intellectual history has done more to confirm Confucius' famous dictum: "When names are not correct, discourse is difficult."[130]

Qu Qiubai never said what he meant by *jinhua lun*. Lu Xun said only that belief in *jinhua lun* meant belief in the supremacy of the young and "the future." Whatever he meant by that, however, his contention that before 1927 he had "only believed in *jinhua lun*," and thereafter his belief had been "blown to bits," was utter nonsense.

Assume first that Lu Xun meant by *jinhua lun* Darwin's theory of evolution. Lu Xun believed in Darwin's theory of evolution from the day he read *Tianyan lun* until the day he died. As we have seen, he did not have a deep understanding of Darwin's theory, but he never doubted that biological evolution was a fact. In 1930, in an article in which he ridiculed Liang Shiqiu's use of Darwinian idiom to declare the proletariat unfit, he still used the ultimate acceptance of Darwin's theory, over religious opposition, as proof that the truth will out.[131] In 1933 he said, "Darwin's discovery that living things are evolving let us know that our distant ancestors were related to monkeys. . . . But among all those monkeys' relatives, Darwin, one must admit, was great."[132] In 1935 he said, perfectly simply, "Man is a link in the great chain of evolution."[133] "In short, as Zhang Zhuo has put it, arguing the same point, "Lu Xun did not throw away the theory of evolution; he did not have to, nor should he have."[134]

But if he did not throw away Darwinism, what "biased belief" in *jinhua lun* did he throw away?

As Zhang Zhuo also noted, again most correctly, "Lu Xun often used *jinhua*, development, advance, and progress as a set of interchangeable terms."[135] But did that mean that in accepting the theory of class struggle Lu Xun rejected the theory of progress? Of course not. In 1933 he said "the age is progressing. Steamships and airplanes are everywhere." He said, "History will never back up. Men of letters need not be pessimistic."[136] In 1935 he said, "Cultural reform is like the flowing of a great river. There is no stopping it. . . . Nothing goes back to the old track, it will definitely move on; and nothing maintains its present form, there will definitely be change."[137] Lu Xun echoed Darwin ("Not one living species will transmit its unaltered likeness to a distant futurity"[138]); Lu Xun was echoed by the Chairman ("The human race is constantly developing. The natural world too is constantly developing. . . . All theories of stopping, of pessimism, of helplessness or proud self-satisfaction are erroneous—because these theories are not in accord with the historical fact of some one million years of human social development. Nor are they in accord with what we know so far of the world of nature"[139]). Confusing, as always, the notions of evolution and progress in *jinhua lun,* Lu Xun rejected neither. So what was "blown to bits?"

If he did not reject "true Darwinism" for Marxism, did he reject "Social Darwinism?"

"The establishment of a Marxist world view," said Zhang Zhuo, "in no way demands the abandonment of an evolutionary view."[140] For "Marxists believe," said Yi Zhuxian, "that Darwin's scientific theory of evolution is not only not at odds with Marxism or incompatible with it, 'it can be used as the basis in natural science of the class struggle in history.'" That, as we have seen, was Marx's view. Lenin's view, as dozens of Chinese scholars have refused to let us forget, was that "Marx's dialectic is the newest scientific theory of evolution."[141] But Marx and Lenin never distinguished themselves less as scientists than when they made such claims. How could their hopelessly muddled mixture of idealism, materialism, determinism, voluntarism, moral indignation, moral exhortation, and faith in the inevitable triumph of the forces of history, "independent of human will,"[142] over the evil of human alienation, have a basis in Darwinian science—unless such thought is itself the result of random mutations in the brain, waiting to meet the (quite unpredictable) test of natural selection? Marx and Lenin, Spencer and Sumner, all were Social Darwinists—and so was Lu Xun, if he believed that Marx and Darwin were fellow scientists.

Chinese scholars, however, insist that Lu Xun was not a Social Darwinist, because he "stood on the side of the weak."[143] "It was precisely because Lu Xun firmly upheld the revolutionary position of opposing imperialism and feudalism," said Zhang Zhuo, "that in analyzing society from the point of view of evolution, development, and struggle, he avoided falling into the two dangerous mud holes of popular evolutionism and Social Darwinism."[144]

It is perfectly true, as we have seen, that Lu Xun despised Social Darwinian defenses of imperialism and feudalism, but was his verbal championing of "the weak" scientifically Darwinian? I have tried to prove above that his opposition to imperialism was idealistic and voluntaristic. Evolutionarily, Lu Xun's crusade is difficult to explain. One might argue that he was biologically programmed to support his race (if one can assume that exhortations to underdogs to become top dogs are somehow perfectly natural—and potentially effective), but why within that race should he struggle for the extinction of his class? Difficult to explain in Marxist terms, in Darwinian terms his self confessed hatred of his class[145] is *extremely* difficult to explain. If, with the eye of a Daoist Darwinian, he recognized that "the weak" were really the strong, and shrewdly chose, not being able to lick 'em, to join 'em, he might "justify" his behavior, and theirs. But why would such justification be any different from an imperialist's? Any "strategy" in a storm. All "stands" were for survival.

In any case, Lu Xun "stood on the side of the weak" long before 1927, so we still do not know what biased view of evolution that year's setback to the weak led him to abandon.

Was it that evolutionary progress was slow, steady, natural, and inevitable—without struggle? No. Lu Xun said in 1927, causing endless confusion, "I had always believed in the theory of evolution; I had always thought that the future must surpass the past, that the young must surpass the old," but nowhere do we find in his works before 1927 any dominant faith in peaceful evolution. There was talk of struggle, resistance, indeed bloodshed,[146] from the beginning. The one great cry of hope in peaceful evolution was the madman's cry to all people-eaters, "You can change!" But they did not change. It was the madman who changed—into a people-eater. And the young did not surpass the old. They learned to eat the old—and each other. Ah Q was beaten by both rich and poor. He himself managed to beat Xiao D. And "Xiao D," said Lu Xun, meant "the young are the same."[147] Only rarely did Lu Xun cry out "The future is bright." More often than not in his stories, the forces of darkness reasserted themselves over the forces of light.

So what in evolution's name did Lu Xun mean? The remaining possibility is the most likely. He probably meant that he used to believe that the struggle for existence in human affairs was between races and, within races, between generations, but the Guomindang massacre of the Communists, and a bit of Communist writing, had taught him that the struggle was actually between classes.

If that belief marked a conversion, however, how significant a conversion was it? Again the matter is confused and confusing.

Lu Xun's anti-Haeckelian Haeckelian argument that imperialists were a breed apart was an early example of "Darwinian" class analysis. Much more common, however, was his supraclass analysis of the *guominxing*, the "national nature" or "character" that was the common cause of Chinese unfitness—and "inhumanity." In Lu Xun's stories people of *all* classes ate people.

After 1927 Lu Xun did indeed, in a few famous lines, make a Darwinesque defense of the proletariat. In ridiculing Liang Shiqiu's contention that the proletariat would prove unfit, he implied, at least, the opposite.[148] And once (just once) he said it outright:

> I already hated my own class, which I know so well, and I had no regrets about its demise. Later, reality too taught me it is true that only the newly arisen proletariat has a future.[149]

Was that a prediction of the genetic extinction of his class? Lu Xun never called for that much dog beating. Lu Xun's assertion still made little Darwinian sense. The masses were the masses—the fit—in or out of power. Marx had never said it was biology that made one a member of the bourgeoisie. But again in any case, did Lu Xun's occasionally exercised class analysis mean that he had abandoned his belief in the common unfitness of the *guominxing*? *No.* Two weeks before he died he said, "To this day I still wish someone would translate Smith's *Chinese Characteristics.*" He wanted Chinese to read Arthur Smith's archly humorous, dreadfully prejudiced and, needless to say, most unflattering analysis of alleged general Chinese racial or cultural (it was hard to say which) weaknesses. Lu Xun hoped that "having read of these things, we will examine ourselves, analyze ourselves, decide which of his points are correct, and change, struggle and do something ourselves, without seeking the forgiveness or praise of others, to prove what Chinese are really like."[150]

Was that a return to the cry of the madman, "We can change," a return to the classless Confucian task of "self-cultivation"? Or had Lu Xun never ceased to be a Confucian?

In 1981 some Chinese scholars argued that the more Lu Xun talked of class struggle the less he talked of "the national character"—and that eventually the Marxist theory of class replaced his notion of a national character.[151] I stand with those who argue that Lu Xun's belief in a national character and his wish to change that national character formed a lasting element central to his thought from the days of his first discussions with Xu Shoushang in Tokyo to the day of his death in Shanghai.[152]

How much did Lu Xun's thought really evolve?

In 1981 Lu Xun scholars came up with some eight different peri-odization schemes to explain the evolution of Lu Xun's thought. They all agreed that the story was the story of Lu Xun's conversion to Marxism. They disagreed about the timing of that conversion.

I wish to "blow to bits" that whole debate—with a periodization scheme of my own designed to prove that periodization schemes are not important to our understanding of Lu Xun's thought.

My periodization scheme is a "perfect" one: four periods of nine years each—in which we do *not* see great change, the first beginning in 1900, the second in 1909, the third in 1918, the fourth in 1927: thirty-six years ending sadly, but neatly in 1936.

In 1900 Lu Xun wrote his first recorded poems, "Three poems of Farewell to my Brothers," which mark symbolically the beginning of his "lonely" life as a writer. In this first period, which we have looked at in some detail, he studied in Japan, but gave up his study of mining, and medicine, for literature.

In 1909 he returned to China—and stopped writing. He taught in two middle schools, welcomed the Republican revolution, worked in the "Republic's" Education Ministry, even throughout the misrule of Yuan Shikai, did a bit of translation and esoteric scholarship—and kept quiet.

In 1918 he found his true voice. He wrote "A Madman's Diary" in a vernacular language never heard before, launching the period of his greatest creativity; he wrote most of his short stories, and prose-poems; he created the *zawen*. He also taught in a series of universities, gave speeches, supported student movements, left Beijing to be safe, first for Xiamen University, which he disliked, and then for Zhongshan University in Guangzhou—where Xu Guangping was. He served as professor and dean until the Nationalist Party purge, then quit, ending forever his career in academe.

In 1927 he retreated to Shanghai, whence he dedicated the last nine years of his life mainly to *zawen* wars with intellectual supporters of the Guomindang, and with occasionally antagonistic leftists. He served as nominal head of the League of Leftist Writers, kept working on transla-tions, and encouraged young writers and young practitioners of the new art of woodcuts of social criticism.

Politically what did he do in these periods?

In the first he supported, somewhat, from the sidelines, the Republi-can revolutionaries.

In the second he worked for "the Republicans" (in the Ministry of Education), even after "the Republic" was taken over by the would-be dictator Yuan Shikai. Lu Xun did not criticize Yuan Shikai until Yuan Shikai was dead, but we know from his later short stories and *zawen* that Lu Xun concluded very early that the revolution had not succeeded.

In the third period he expressed that opinion, with devastating satire, voiced his disgust at the warlords' brutal methods of repression, and eventually supported, from the sidelines, the Nationalist Revolution.

In the fourth, shocked by the Nationalists' brutality, and convinced again that the revolution had failed, he supported, from the sidelines, the Communist Revolution.

Progression or pattern? An ascent to enlightenment or a sorry cycle of hope and disappointment—and of sympathy, support, disillusionment, and resistance?

How much evolution do we see in his thought?

Did he slowly but surely evolve into a materialist?

No. He was a would-be materialist from the beginning, and an idealist to the end.

Did he evolve from a peaceful evolutionist into a realistic revolutionist?

No. He never was a peaceful evolutionist. Evolution, he said from the beginning, taught struggle and resistance—even though the struggle he usually spoke of was intellectual (his "daggers and spears" were essays). His message was always ambiguous. He decried bloodshed, but prophesied that nothing could be changed without bloodshed. His rhetoric was militant from the beginning, but to the end he shied away from urging young people to take up arms.

Did he evolve into a great Marxist?

No. Within the severe limitations set by Guomindang censors (and assassins) he did his best to ridicule redbaiters. He defended the Soviet Union (quite uncritically[153]). He read some Marxist books, chose a Japanese introduction to Marxism as a Japanese textbook for his wife, helped translate works on Marxist theories of literature, and in a *few* articles, laden with Marxist language, defended those theories, and propagated them.[154] But that was all. Except for some clever rhetoric, he made no contributions to Marxist theory. Of Marxist philosophic and economic theory he made almost no mention whatsoever. In all, of Marxism he said very little. In his final period he attacked unceasingly supporters of the Guomindang and praised Marxism in here an essay there an essay—just enough so the Communists could say, when he died, "He was on our side."

But would he have stayed there—or would "the cycle" have borne him once again into the ranks of the opposition?

Had he lived, I think for even one more period, we would know. Is it idle to speculate? In all four of our periods and all three of his causes, Lu

Xun was "lonely"—independent, and critical. He sympathized with three parties. He actively joined none. He was not a party animal. He took no party orders. He was always an outsider, always critical of enemies *and* comrades.[155] He criticized the republican martyr Qiu Jin in his first period and (much more harshly) the Communists Zhou Yang, Tian Han, Xia Yan, Yang Hansheng, and Xu Maoyong in his fourth.

Had he lived, is it imaginable that the cycle would not have started round again? He defended Hu Feng in 1936. Would he not have defended Ding Ling, Xiao Jun, Wang Shiwei, and the rest in 1942? Could he have stopped writing *zawen* himself? Would he have accepted *all* the violence of the Revolution as necessary dog beating? Would he have sat silent through thought reform? Would he have joined in the criticism of Feng Xuefeng in 1954 and Hu Feng in 1955, as his wife did? Would he have ignored the Chairman's call for a Hundred Flowers in 1956? Would his courage have failed him in 1957?

It is not only foreign infidels who speculate. Here is the full text of Zhang Yu'an's 1980 poem cited above, "Jiaru ta hai huozhe" (If he were still living):

> If he were still living, I do not know
> What people would call him.
> If he were still living, I do not know
> What he would urge us to do.
>
> Perhaps he would hold high position
> But perhaps—he would be only a soldier.
> In high office—he would not forget his promise to be an ox
>   for the young.
> In low estate—he would not act the fawning slave!
>
> Perhaps he would already have received many honors,
> But perhaps—he would just have been let out of jail.
> Honored, he would cry out and pace back and forth anew.
> In jail, he would rewrite his *Permitted Discussions of Wind and
>   Moon* and *On False Liberty*. . . .
>
> Perhaps he would no longer carry his notes in that patterned
>   paper bag.
> But surely he would not disdainfully walk about with his
>   nose in the air.
> Perhaps he would attend important meetings,
> But not followed by two secretaries and three bodyguards.

Perhaps he would ride in a modern sedan,
But surely he would not use curtains to shut out the outside.
He would reach out to the destitute,
He would quietly read the complaints of the many young
      still waiting to be employed.

Perhaps he would ever be spilling ink in hymns to "the new
      life,"
But perhaps—he would be lancing with his pen the ills of
      the age.
Perhaps he would enjoy more joy and laughter,
But perhaps—he would feel new uneasiness and rage.[156]

Poets have more license than historians. Perhaps historians have indeed no right to those "perhapses." But in them, in the insistent
speculation of people like Zhang Yu'an that Lu Xun would have turned, in
turn, into a critic of the Communists, there lies a clue to the answer to our
question, "Does 'the evolution of Lu Xun' reveal 'the real Lu Xun?'" for
Zhang Yu'an's speculation, his attack on the official Lu Xun, was based on
something real in the writings of Lu Xun, on something to which Wang
Hui, too, obliquely pointed, in his "Historical Criticism of Lu Xun Research," something real and unchanging—throughout all four of our
unimportant periods.

We have only clues. Zhang Yu'an and Wang Hui did not make
themselves perfectly clear. They were not in any case writing strictly academic analyses of the real Lu Xun. They were like Lu Xun, through Lu
Xun, lancing evils of the day, Zhang Yu'an aiming his pen at officialdom in
general, Wang Hui aiming his at politics in command—of scholarship,
ever mindful, however, that politics were still enough in command to
make a certain lack of clarity the better part of valor. So we have only
clues—but useful clues.

Look at two final points in Wang Hui's critique. Wang Hui criticized his colleagues (most specifically Wang Furen, whose watchword
"Back to Lu Xun"—the real Lu Xun—he nonetheless heartily applauded)[157] for failing to see in Lu Xun's thought "a ubiquitous 'cyclical
feeling,' a 'sense of repetition,' or an experience of 'transmigration.'" To
Lu Xun, he said, "the process of history seemed nothing but a succession of repetitions and cycles time after time, and present reality—
including the movements he had personally experienced, seemed to
mark no '*jinhua,*' or progress, but only an absurd transmigration." In
short, he said, "Whether at a conscious, rational level or an unconscious,
emotional level, there was always in Lu Xun's mind a kind of anti-evolu-

tionary logic [Wang Hui too confused evolution and progress] and emotional tendency."[158]

We have already seen this, many times over, in Lu Xun's early parable of the iron room, in his madman's "recovery," in his portraits of revolting revolutionaries,[159] in his conviction that "the banquet" still went on, in his fear that he wrote of the future not of the past, in his sense, even after his "conversion," of the tenacity of "Chinese characteristics."[160]

Wang Hui said that Lu Xun "constantly discovered in the people of his time, in his friends, even in the persons of his comrades in arms that 'the past' was not yet past."[161]

That, of course, is what Zhang Yu'an meant when he said Lu Xun might still find cause for rage, when he wove the titles of so many of Lu Xun's books into his poem to show that what Lu Xun said he still could say.

That is what all those Lu Xun scholars meant who said Lu Xun's work was of "practical significance."

That, to be fair, is what Chinese leaders meant, when they admitted that "feudalism" was not yet dead.

On what, however, was Lu Xun's sense that "the past was not yet past" based? On his conversion to Marxism? No. On some traditional belief in cosmic cycles? No. It was based on his *judgment* of his *experience*, and *that* was based on an unchanged, unchanging, moral sense.

Wang Hui also criticized Wang Furen, and other colleagues, for falling prey, even in their laudably revisionist studies of Lu Xun, to theories of "necessity"—to determinism, for failing to escape from the determinism of earlier Lu Xun scholars like Chen Yong, who once insisted that Lu Xun's evolution "from Democracy to Communism" was "determined from the very first day that Lu Xun entered his life in Chinese literature."[162] The notion that Lu Xun's conversion to Communism was inevitable, taken together with Mao Zedong's dogma that Lu Xun's "direction" was the direction of modern Chinese culture, was most comforting to the faithful. It ignored, however, said Wang Hui, the twentieth century's all important discovery of "chance."[163]

What did that mean? He did not say exactly. But what might it mean to say that Lu Xun turned "pro Communist" not "inevitably" but "per chance?" What if Lu Xun was drawn to the Communists not because they held magnetic truth, but because he was against the Guomindang and the Communists "chanced" to be the only opposition in the field?

Lu Xun had sympathized with "the opposition" twice before. Why? Because he hated people-eating people. Imperialists ate people, the Chinese imperial system ate people, the Chinese tradition ate people, successful revolutionaries ate people, warlords and their running dogs ate people, the Nationalists and their running dogs ate people—and that is why people think Lu Xun would in time have opposed the Communists, because they ate people.

In all of that do we see inevitable progressive evolution, a simple cycle of discontent—or something constant?

The Real Lu Xun

We have already seen "the real Lu Xun"—not the real "person," not Lu Xun the son, brother, husband, father, teacher, colleague, comrade, friend, but Lu Xun the *rendaozhuyizhe*, the humanist or humanitarian, an "imperfect" humanitarian, but a "perfectly human" humanitarian, flawed only in his "righteous indignation" that justified, rhetorically, inhumanity to the inhumane.[164]

The myth of Lu Xun's evolution has obscured the real Lu Xun—in three ways.

First, it has simply silenced him in general, by saying that what he really had to say was simply that the Communists were right—and that once one realized that, one need pay him no heed.

Second, the myth has protected the Communists from his criticism, by tying the objects of his criticism to history: With ever-increasing clarity, so the makers of the myth have said, Lu Xun attacked the enemies *of his day*, finally realizing that China's enemies were class enemies and so, obviously, "not us."

Third, by insisting that the truth lay in change, the myth has obscured the truth that the heart and soul of Lu Xun's thought did *not* change.

The first two obscurations Wang Hui denounced in 1988, with startling (if turgid) outspokenness:

From the angle of political ideology, to simplify the path of Lu Xun's life and the course of his mental development and call it the "basic direction" and the "basic law" of progression from democracy to communism, is to say that the basic mission of Lu Xun research is to prove true this sacred and absolute "direction" and "law." Lu Xun—who did not want to prove the permanence of anything, but untiringly exposed the transience of present existence—has become himself something "absolute" and "sacred." Out of fear of the new content of reform, produced as socialism is transformed from an idea into a reality, and out of fear that Lu Xun's spirit of stubborn and inflexible criticism and denunciation might reveal the limitations and impermanence of present life and become a source of power to encourage change of those limitations, Lu Xun's critical and denunciatory style of writing has been declared outmoded and unfit for our new era. Lu Xun's tradition of realism, as a literary proof or "mirror" of the progress of China's democratic revolution, is "clear-headed" and "thorough," but it belongs to history, it belongs to the past, it belongs to a critical summary of an earlier period; it is no longer itself a force in life, no longer an immortal force for revolutionary change.[165]

Wang Hui saw that the Communists had done to Lu Xun what Levenson said twenty years ago they had done to Confucius: They had "pushed [him] back in history" and "museumified" him. "The 'museumified' Confucius," said Levenson, "does not speak."[166] The "museumified" Lu Xun, said Wang Hui, does not speak either.

Lu Xun was right: writers could be "praised to death."[167] The Chairman was right: "Some people, who defend Lu Xun with their every utterance, actually go against him."[168] The Chairman himself had led the way. Lu Xun scholars had followed, and Lu Xun research had "lost its significance and its value."[169]

That was Wang Hui's angry and discouraged charge. He saw how the myth of Lu Xun's evolution had helped silence Lu Xun *in general*. He did not see—or choose to describe—how the myth specifically obscured Lu Xun's undying hatred of people eating.

But Lu Xun, in any case, could not be silenced. Wang Hui's angry, discouraged charge was exaggerated. Rumors of Lu Xun's death by praise were exaggerated. Wang Hui, Zhang Yu'an, and all the Lu Xun scholars who kept saying that Lu Xun had much to say proved it—even when they did not dare say what he had to say.

Wang Junji reported after a Lu Xun conference in 1980:

Some comrades have said that young people have a very deep affection for Lu Xun's works. "The Gang of Four" wanted to distort and use Lu Xun, but many people studying Lu Xun's works in the Decade of Disaster raised their ability to see through "the Gang of Four." In the fall of "the Gang of Four," Lu Xun's works played a positive role. In the Tiananmen Incident many young people drew wisdom and strength from Lu Xun's works.[170]

If Lu Xun spoke to the first Tiananmen Incident, in 1976, would he have been silent after the second, in 1989? Could he have been *kept* silent, in "the Museum of History," that stares across Tiananmen Square at "the Great Hall of the People?"

Thumb through Lu Xun's works and you can hear him speak:

What I fear is that my random essays still seem to speak of the present, or even of next year.[171]

> The tears have been wiped away,
>   The blood has been washed away;
>   And the butchers go free and take their ease.[172]

China is especially ferocious towards its own.[173]

The butchers of the present . . . have killed "the present" and so have killed "the future"—and the future is our children's time and our grandchildren's.[174]

In my experience, most of those who pose as "revolutionaries" and cavalierly denounce others as "traitors to the party," or "counter-revolutionaries," or "Trotskyites," or even "traitors to the race," are not people on the right road; because they are subtly killing the revolutionary power of the people, they are not looking out for the interests of the revolutionary masses, they are just using the revolution for private gain.[175]

Actually, those who now hold power are the so called revolutionary youth of yesteryear.[176]

They have long since taken care of those with spirit, killing or imprisoning them.[177]

Politicians identify writers as agitators, who will disrupt society. They think that if they kill them, society can be kept peaceful. They do not realize that even if they kill the writers, society will still revolt.[178]

Waves of irony. The leaders of the People's Republic became twice over their own worst enemies (the process began long before 1989). They turned into the warlords and nationalists of the past, surpassing them in the enormity of their repression. And they threatened their own existence, creating through their acts what they feared most: the threat of "counter-revolution."

Lu Xun stood up from his works and accused them: "You eat people. Even in your *'renyi daode,'* in your 'morality' you eat people."

But then what? What did he tell the new "New Youth" to do? The *Communists* told them he told them to rebel. The *Communists* told them he told them to show people-eaters no mercy. The *Communists* told them he was for revenge and against forgiveness, he hated "humanitarians," he knew that people-eaters never exit on their own, they never fall unless you hit them.

So if the new New Youth studied the Communists' Lu Xun, the Communists were doomed.

The Communists lionized Lu Xun, forgetting—a thousand pardons—that "to raise a tiger is to court disaster."[179]

Was "the real Lu Xun" really so ruthless?

There is no simple answer, because the real Lu Xun contradicted himself. We come back to "the great contradiction": Beat dogs but do not eat people. Lu Xun said both, but we must choose. And China's new youth must choose. "Choosing we choose, not choosing we choose."[180] We are condemned to choose.

We must choose between Confucius' two ways, "the humane and the inhumane." There is no third way.

Lu Xun thought he walked a middle path, as we all do, a muddy middle path—but there is none. We all slide back and forth between Confucius' two ways. The postponement of fair play is no way out. Postponement is a choice.

But we must *see* that it is no way out. We must see the contradiction before we choose. And that is why the great contradiction in Lu Xun's thought is a great gift: It shows us our choice in all its starkness—and helps us choose humanity.

For Lu Xun shows us inhumanity in the limitations of *his* humanity, and inhumanity is what he set out to show us in the first place.

Lu Xun is great because he shows us inhumanity. He is great because he shows us—all of us—that we eat people. So it is not surprising that we make out the words "eat people" not only between the lines of the madman's Chinese chronicles, but between some of Lu Xun's own lines as well. Those are the words he prepared us to see, even when he did not see them himself. Unwittingly showing us the limitations of his humanity, he helps us recognize the limitations of our own.

Lu Xun wrote about Chinese, because he was most worried about Chinese, but he is great because, without trying to, he wrote about us all.

The dog in "The Dog's Retort," the one dog in all Lu Xun's works who puts people in their place, puts all people in their place:

I dreamt that I was walking in a narrow lane, wearing tattered clothes and worn out shoes, looking like a beggar.

A dog started barking behind my back.

I haughtily turned my head and cursed him saying, "Hey! Shut up! You snobbish cur!"

"Hee, hee," he laughed, and then said: "I dare not so presume. To my shame I'm no match for you people."

"What!?" I was furious, thinking this an intolerable insult.

"I'm ashamed to say I still can't distinguish copper and silver; I still can't distinguish cotton and silk; I still can't distinguish officials and people; I still can't distinguish masters and slaves; I still can't distinguish—"

I fled.

"But wait! Let us talk some more . . . :" He loudly tried to hold me back.

I fled all the way, as fast as I could, until I fled out of my dream, and back to my bed.[181]

Lu Xun wrote to wake us up. That is what he confessed in the preface to his first collection of short stories, his "cry." He cried out to wake us up, even though he feared we were all prisoners in an iron room—and he with us. He cried out to wake us up to our inhumanity.

Later we can see inhumanity in some of his own angry cries, but that is hardly odd, as he never claimed to be above us. But his humanity came before his inhumanity. It was there first, and it was always there, beneath it. It had to be, because he had to love before he could hate. He had to love people before he could hate people who ate people.

So the *real* Lu Xun was a true Confucian.

Where does that leave evolution? Not even touching the real Lu Xun, in any way we can comprehend. Unless "the real Lu Xun" is unreal.

The idea of evolution fascinated Lu Xun, it excited Lu Xun, it encouraged Lu Xun, it inspired Lu Xun, it (sometimes) depressed Lu Xun. Did it not "influence" him?

He said it did. Hundreds of Chinese scholars have said it did. But the harder we look at that influence, the less substantial it appears, and the *real* Lu Xun appears only when we see through it. And so we must see through it.

Look at what evolution first meant to Lu Xun. It excited him in two ways, as it did Yan Fu and Liang Qichao, as something totally new and non-Chinese, as something that opened up worlds beyond the Chinese tradition, and conversely as something that might save China. It spoke to Lu Xun as someone who was *already* a young nationalist, who had dedicated himself to struggle for his *nation's* survival. (That is why he was studying mining.) That is why, despite his honest interest in natural history, he immediately seized on Darwinism as a *social* theory (which it was *not*). It told him that China must struggle to survive, it must seize control of its own resources, and it must *change*. That was the magic word, as it was to Liang Qichao, because Lu Xun *wanted* change, because he did not *like* China, though he loved it, because he was convinced, before he ever heard of Darwin, that China must change to survive. Darwin made change "scientific," so Lu Xun, like Yan Fu, Liang Qichao, Sun Yatsen and everybody else, looked to Darwin's theory of evolution for the secret of change, the secret of national fitness.

But then Lu Xun changed. He distinguished himself by losing interest in the forms of fitness that other nationalists espoused, physical, military, institutional, economic. He rededicated his life to the pursuit of Chinese moral fitness—the quintessential Confucian Way to save the state, although he attacked Confucianism as he pursued the Confucian Way. He distinguished himself too by becoming more than a nationalist, as a good Confucian should, caring not first and foremost for the survival of the state, but "for the people." Not for all of the people. This is where he limited his Confucian "humanity." He said that incorrigible people-eaters should be beaten, and he sometimes implied they were a class, the ruling class that survived each revolution. But that, as we have said, was not odd. All of China's political groups dehumanized their enemies, Manchus, republicans, warlords, Nationalists, Communists, imperialists. What was odd was that Lu Xun did not start or stop with any dehumanizing scapegoat class. He saw the ultimate enemy in us ourselves. And he tried to wake us up to our evil so that we would change ourselves.

It is true he almost despaired of our changing. All his life he wrestled with despair. It was not just once that he felt we were in an iron room. But the madman's Confucian cry, "You can change," day by day beat back despair—for Lu Xun kept writing.

In his struggle he used Darwin's theory of evolution. He used it on both sides of his struggle with himself and with his enemies. He used Darwin's theory of evolution to condemn those with "animal natures" and to prophesy the evolution of "true people," "humane people," Confucian people. He used Darwin's theory of evolution to condemn people-eating and to justify dog-beating. He used Darwin's theory of evolution, as a hundred Chinese scholars have said, "as a weapon," but as a weapon in the struggle for "justice" (again in the Greek sense) in a purely idealist struggle, not a materialist struggle, not a Darwinian struggle, in a struggle that ultimately makes no Darwinian sense.

And we have seen that in that struggle, Lu Xun's evolutionary arguments made little Darwinian sense. It was not just his great contradiction that weakened his arguments. We have seen that he was not scientific on either side of his great contradiction. But in any case, the great contradiction was only a contradiction over the human way to justice. There was no contradiction over Lu Xun's certainty that justice was the Way. Justice, goodness, *ren*—that, for *ren*[182] was the only Way.

Did Lu Xun learn that from evolution? Of course not. Might he have, if his knowledge of evolution had been more scientific? No. Our knowledge of evolution teaches us nothing of the sort. "The teachings of the naturalist respecting," in Huxley's peculiar phrase, "that great Alps and Andes of the living world—Man,"[183] have taught us nothing about what men, women and children *should do,* or of why we feel we *should do* anything. "The teachings of the naturalist" can only attempt to explain our feelings away.

Great light *has been* "thrown on the origin of man and his history,"[184] but we are still in the dark. Whole libraries have been added to our knowledge of natural history, but "morality" is still a mystery—or nothing.

But we read Lu Xun. All sorts of people read Lu Xun, because we feel Lu Xun has lightened our darkness—by showing us our darkness.

But how could he see in the dark? How do we see in the dark?

We see, if we see, by reflected light.

"Tonight there is good moonlight," the madman said.[185]

Lu Xun was a madman, or a mutant, or in Huxley's even more peculiar phrase, a man, "reflecting, here and there, a ray from the infinite source of truth."[186]

Abbreviations Used in Notes

CCD   James Pusey, *China and Charles Darwin*
EE   Thomas H. Huxley, *Evolution and Ethics*
EM   Ernst Haeckel, *The Evolution of Man*
HC   Ernst Haeckel, *The History of Creation*
LX   *Lu Xun quan ji*
LXYJ   *Lu Xun yanjiu*
LXYJZL *Lu Xun yanjiu ziliao*
M   Ernst Haeckel, *Monism*
O   Charles Darwin, *On the Origin of Species*
R   Ernst Haeckel, *The Riddle of the Universe*

To find quotations from Lu Xan in editions of his works other than that cited, the 1981 edition of *Lu Xan quan ji* (The complete works of Lu Xun), use the following breakdown of the contents of the 16 volumes of that edition—and a calculator:

LX, 1: 3 *Fen;* 291 *Re Feng;* 415–570 *Nahan.*
LX, 2: 5 *Panghuang;* 159 *Ye cao;* 229 *Zhao hua xi shi;* 341–481 *Gu shi xin bian.*
LX, 3: 3 *Hua gai ji;* 183 *Hua gai ji xubian;* 407–583 *Er yi ji.*
LX, 4: 3 *San xian ji;* 189 *Er xin ji;* 417–653 *Nan qiang bei diao ji.*
LX, 5: 3 *Wei ziyou shu;* 189 *Zhun feng yue tan;* 417–590 *Hua bian wenxue.*
LX, 6: 3 *Qie jie ting zawen;* 217 *Qie jie ting zawen er ji;* 469–637 *Qie jie ting zawen mobian.*
LX, 7: 3 *Ji wai ji;* 257–466 *Ji wai ji shiyi.*
LX, 8: 3–479 *Ji wai ji shiyi bubian.*
LX, 9: 1 *Zhongguo xiaoshuo shilue;* 341–426 *Han wenxue shi gangyao.*
LX, 10: 3 *Guji xuba ji.* 151–479 *Yiwen xuba ji.*
LX, 11: 3 *Liang di shu;* 319–699 Letters (1904–1929).
LX, 12: Letters (1930–1934) 1 1930; 33 1931; 67 1932; 137 1933; 313–630 1934.
LX, 13: Letters (1935–1936) 1 1935; 283 1936; 453 To foreigners; 677–690 msc.
LX, 14: Diary (1912–1931) 1–907.
LX, 15: Diary (1932–1936) 1–317.
LX, 16: Indices.

Notes

A Necessary Preface

1. Wu Han, *Chuntian ji* (Beijing, 1961), p. 28 (see also pp. 35, 178, 200, 137) and *Deng xia ji* (Beijing, 1957), p. 168. Wu Han (see below, p. 143) was a historian famed in the thirties and forties for veiled criticism, through historical vignettes, of the Guomindang (The Nationalist Party, the Kuomintang in Wade-Giles romanization, hence, in earlier works, and common parlance, the KMT). He tried the same technique in the People's Republic after the Great Leap Forward. His criticism was not appreciated by Chairman Mao. He was denounced in the prelude to the Cultural Revolution, and became one of that revolution's first victims. See James R. Pusey, *Wu Han: Attacking the Present Through the Past* (Cambridge, 1969). When Wu Han said that a person was "worth our study today," he often meant worth our "imitation," which is not exactly what I mean here.

2. *He er er yi*—I borrow, without ideological intent, one of a highly charged pair of phrases that appeared in the title of an article published in *Guangming ribao* (The Guangming Daily) on May 25, 1964, " 'Yi fen wei er' yu 'he er er yi'" ("One divides in two" and "two join and become one"). Allegedly intended to initiate a philosophical discussion of the theory of the unity of opposites, that "academic" article launched instead a political "debate" that became part of the "intellectual" prelude to the Great Proletarian Cultural Revolution (1966–1969), or, indeed, to what the Communist Party itself now calls "the Decade of Disaster (1966–1976), the decade in which Mao Zedong fought to regain control of "his" revolution. *He er er yi* was said to represent the counterrevolutionary intent of those who favored class unity over class struggle. See *Zhongguo dabaike quanshu, zhexue* (Beijing, Shanghai, 1987), vol. II, p. 1072.

3. O, p. 484.

4. Ibid., p. 488.

5. Thomas H. Huxley, *Man's Place in Nature* (Ann Arbor, 1959), p. 71 and title.

6. *Zhuang Zi ji shi* (Beijing, 1961), Chapter 17, vol. 3, p. 584. The Daoist (Taoist) philosopher, Zhuang Zi (Chuang Tzu) lets a very perplexed River Lord ask this question near the end of a Daoist sermonette designed to destroy distinctions.

7. See *CCD*.

8. See Ibid. "May Fourthists," a term to be used only this once, in desperate want of a better, refers specifically to students who took part in the great demonstration on May 4, 1919, to protest the acceptance at the Versaille Peace Conference by England, France, Italy, and, alas, the United States, of Japan's de facto occupation of most of Shandong Province, seized from Germany during the war. Generally, the term can include all those intellectuals who took part in the New Culture Movement from 1915 to 1921.

9. Chinese Social Darwinism was no stranger than anyone else's Social Darwinism. The ubiquitous ease with which peoples round the world have leapt to un-Darwinian conclusions about Darwinism (as Darwin, on occasion, did himself) offers yet more evidence of racial equality. If Social Darwinists, however, are all of a breed, they are not of a creed. I use the term Social Darwinism, therefore, only in a most general sense and affix the (consequently not very useful) label Social Darwinist as I think it should be affixed, not just to capitalists, imperialists and white supremists, but to all who find "scientific sanctions" for social action, or inaction, in the works of Charles Darwin.

10. Lu Xun's good friend, Xu Shoushang, wrote in a recollection of unspecified time, but probably of their student days in Japan: "One day we happened to talk of *Tianyan lun*, and Lu Xun could recite many chapters by heart. And I? Actually, I could recite several chapters too. So suddenly the two of us started to recite Chapter One. . . ." See Xu Shoushang, *Wang you Lu Xun yinxiang ji* (Beijing, 1981), p. 8. I have convinced myself that I read in some other well lost memoir that Lu Xun could still do this late in life.

11. [Lu Xun yulu], (no title, n.p.n.d.), pp. 4, 7.

12. For the story of his being "pushed" into writing for *Xin qingnian*, see LX, pp. 418–419. Surely Chen Duxiu and Hu Shi, not Lu Xun, started the "Literary Revolution." That has led Chou Chih-p'ing, a leading Hu Shi scholar, among other things, at Princeton, to mutter with some justification, "Without Hu Shi there would have been no Lu Xun."

13. *Mao Zedong xuanji* (Beijing, 1969), 2.658.

14. The other one, of course, being "the little red book," *Mao zhuxi yulu* (Quotations from Chairman Mao). Because of the resemblance, I have, for convenience, referred to this work as [*Lu Xun yulu*] (quotations from Lu Xun). See bibliography.

15. From the famous title, of course, of Edwin O. Reischauer and John K. Fairbank, *East Asia: The Great Tradition* (Boston, 1960).

16. See below, p. 144.

17. I have been told by several people that they read Lu Xun's complete works in prison, house, or "unit" arrest during the Cultural Revolution.

18. These latter phrases all come from Chinese students' comments to me about Lu Xun. "The Gang of Four," was a title supposedly bestowed by Chairman Mao himself on his wife, Jiang Qing, and his and her Shanghai henchmen, Zhang Chunqiao, Yao Wenyuan, and Wang Hongwen, who helped the Chairman foment and "carry out to the end" the "Great Proletarian Cultural Revolution." When Mao Zedong died, on September 9, 1976, and the Gang was arrested, on October 6, 1976, the Gang of Four was blamed, for some time, for the entire "Decade of Disaster," so that the Party could spare itself the embarrassment of blaming "the helmsman" himself. Before long, however, Westerners visiting China reported conversations with Chinese who when speaking of the Gang of Four, held up five fingers.

19. Quoted by hundreds. See, for example, Hu Sheng, *Zao xia lun cong* (Beijing, 1978), p. 22.

20. LX, 3.379.

21. LX, 1.216.

22. Edward O. Wilson, *Sociobiology: The New Synthesis* (Cambridge, 1975), p. 287.

23. Thomas H. Huxley, *Science and Christian Tradition* (London, 1897), pp. 43–44.

1. A Mentor Once Removed

1. "The Dynasty," China's last, one assumes, was the Qing Dynasty of the Manchus, who ruled China from 1644 to 1911. Lu Xun's school, founded in 1896 as a Railway School by the conservative reformer Zhang Zhidong, was later expanded by Liu Kunyi to become the Nanjing School of Mining and Railways. See William Ayers, *Chang Chih-tung and Educational reform in China* (Cambridge, 1971), p. 106 and Lin Fei and Liu Zaifu, *Lu Xun zhuan* (Beijing, 1981), p. 25. Lu Xun entered the school in January 1899 and graduated in January 1902. He is now said to have bought his copy of *Tianyan lun* in 1901, although a more exact date remains unclear. See *Lu Xun nianpu* (Hefei, 1979), p. 40. *The Origin of Species* could not be sampled in Chinese until 1903, when Ma Junwu published his translation of chapters three and four, "Stuggle for Existence" and "Natural Selection." He published the first five chapters in 1904 and a complete translation, China's first, in 1919. See CCD, p. 306.

2. I use the term "imperialist" not as a derogatory epithet, but as a simple, and accurate, descriptive term for countries with empires, large or small. In 1901, the United States had "acquired" The Philippines and Hawaii, and Japan, Taiwan.

3. LX, 2.295–296.

4. See above, Chapter I, n. 10.

5. For what Yan Fu did with, or to, Huxley's title, see CCD, pp. 172–173.

6. According to William Irvine, "Episcopophagous" (bishop eating) was an epithet Huxley awarded himself. See William Irvine, *Apes, Angels, and Victorians* (New York, 1955), p. 339.

7. *Zhongguo zhexue shi jiaoxue ziliao xuanji*, ed. Beijing Daxue zhexue xi Zhongguo zhexue shi jiaoyan shi (Beijing, 1989), vol. I, p. 182.

8. EE, pp. 82, 115 n., 21, 82.

9. For Huxley's "pigeon fanciers," see EE, p. 23. For Spencer's first sentence, see Richard Hofstadter, *Social Darwinism in American Thought* (Boston, 1955), p. 41. For his second, see Herbert Spencer, *Social Statics* (New York, 1896), p. 151.

10. O, p. 489.

11. EE, p. 85.

12. Ibid., p. 44.

13. Ibid., p. 53.

14. Andrew Carnegie, quoted in Hofstadter, p. 45.

15. Huxley, *Science*), p. 256.

16. EE, p. 71. See also pp. 44, 85.

17. Ibid., p. 79.

18. Edward Livingston Youmans, quoted in Hofstadter, pp. 47–48.

19. EE, p. 3.

20. Ibid., p. 82.

21. Ibid., p. 75.

22. Ibid., p. 81.

23. Ibid., pp. 51–53. Huxley's "ape and tiger" were borrowed from Tennyson, who had introduced them in these lines from his proto-evolutionary poem, "In Memoriam, 118," published in 1850:

> Move upward, working out the beast,
> And let the ape and tiger die.

Huxley, in 1893, could count on his audience catching his allusion. I missed it, for years, until I happily stumbled on the above lines quoted by Gertrude Himmelfarb, in her work, *Marriage and Morals among the Victorians* (New York, 1986), p. 80.

24. O, p. 488.

25. Huxley, *Man's Place*, p. 71.

26. Ibid., p. 131.

27. Ibid., pp. 129–130.

28. Ibid., p. 132.

29. See Charles Darwin, *The Voyage of the Beagle* (New York, 1962), p. 500, and Nora Barlow, ed. *The Autobiography of Charles Darwin* (New York, 1969), p. 91.

30. EE, p. 59.

31. Ibid., p. 11.

32. Ibid., p. 79.

33. Barlow, p. 93, n 2.

34. EE, p. 12.

35. See above, note 11.

36. EE, pp. 85, 45.

37. A mortal saying of Churchy La Femme, lost in the Pogonian canon.

38. See above, p. 3.

39. EE, p. 86. Huxley quotes (out of order) from three couplets from Tennyson's *Ulysses*.

40. "Though you drive Nature out with a pitchfork, it will always return." Huxley quoted this line (from Horace's *Epistles*, Book I, 10) in a footnote that defies rediscovery.

41. Albert Camus, *The Myth of Sisyphus* (New York, 1955), p. 91.

42. EE, p. 44.

43. See above, p. 4.

44. Huxley, *Science*, p. 257.

45. EE, p. 80.

46. Huxley, *Science*, p. 318.

2. The Pen, Not the Scalpel or the Sword

1. Lin, Liu, pp. 47–48.

2. See above, p. xi.

3. See for example Wang Shiqing [*sic*], *Lu Xun zaoqi wu pian lunwen zhuyi* (Tianjin, 1981). Mr. Wang pronounces the last character of his name "qing," although dictionaries insist on "jing."

4. Xu Shoushang, *Wo suo renshi de Lu Xun*, pp. 18–19, cited without place or date in *Lu Xun nianpu* (Hefei, 1979), p. 46. For a similar passage, see Xu, *Wang you*, p. 19.

5. CCD, p. 99.

6. Ibid., p. 209.

7. O, p. 5 (my italics).

8. *Xin min congbao*, Feng Zishan, ed. (Taibei, 1966), vol. IV, 22:3.

9. Ibid., vol. IV, 22:4. See also *CCD*, pp. 190, 209–210.

10. See CCD, pp. 190–192, 209–210.

11. *Lu Xun nianpu* (Hefei), p. 46.

12. LX, 7.12–16. The Latin is from Horace, *Odes*, 3.2.13.

13. See Lin, Liu, p. 41, and CCD, pp. 214, 264. Liang Qichao had tried to inspire a martial spirit in his people at least as early as 1899, when he wrote his essay, "Zhongguo zhi hun an zai hu" (Where Is China's soul?), but he was searching then not yet for Spartan spirit but for something akin to "Japan's soul," *bushido* (the way of the warrior). He had found a model in Sparta, however, at least by the time he published "Lun shang wu" (On revering war) on March 27, 1903. See Liang Qichao, *Ziyou shu* (Taibei, 1960), p. 37, and *Xin min shuo* (Taibei, 1959), pp. 108–109.

14. Lin, Liu, pp. 40–41.

15. LX, 7.9.

16. Ibid., 1.425.

17. Ibid., 7.14.

18. See Lin, Liu, pp. 56–62.

19. Lu Xun said precious little about his marriage, but filial piety was his only excuse for going through with it. To his friend Xu Shoushang he said of his bride: "This is a present from my mother. I can only do my best to take care of it. I know no love" (and he said "it," not "her"). To other friends he said, "My mother has taken a daughter-in-law." Only later did he come up with an excuse both filial and patriotic: "It was then an age of revolution. I could not be sure when I would die. My mother wanted a companion, so I let her have her way." See *Lu Xun nianpu* (Hefei), pp. 66–67.

20. LX, 7.4–5.

21. See Lin, Liu, pp. 39–40, LX, 16.5, and Victor Hugo, *Oeuvres politique complètes—Oeuvres diverses* (Paris, 1964), p. 1654.

22. *De La Terre à la Lune* (1865) and *Voyage au Centre de la Terre* (1864). Lu Xun published a translation of the first two chapters of the latter in December of 1903 and the complete book in March 1906. See LX, 16.6.

23. LX, 10.151–152.

24. Wang Shiqing, *Wu pian*, p. 8.

25. Ibid., p. 2.

26. *Zhongguo kuangchan zhi* (China's mineral resources) and *Zhongguo kuangchan quan tu* (A complete map of China's minerals), compiled with Gu Lang and published in Shanghai in May or July 1906. For slightly varying accounts see LX, 16.6 and *Lu Xun nianpu* (Hefei), p. 65.

27. LX, 8.5, 16.

28. Ibid., 8.11.

29. Ibid., 8.8.

30. Ibid., 8.16.

31. Ibid., 8.8.

32. Ibid., 8.3.

33. Ibid., 8.17.

34. Ibid., 8.3.

35. Ibid., 8.4.

36. Ibid., 8.6.

37. See CCD, pp. 334–355.

38. I have never set hands on the *locus classicus* of this favorite Cultural Revolution quotation from Chairman Mao, but in David Milton, Nancy Milton and Franz Schurman, *People's China* (New York, 1974), p. 239, it is identified as coming "from a speech at the Rally of People of All Walks of Life in Yan'an to Celebrate the Sixtieth Birthday of Stalin," which was in 1939. Oddly enough the passage in question was not included either in the Chairman's selected works or in the famous "little red book" of quotations.

39. Ibid., pp. 190–192.

40. Lin, Liu, p. 35.

41. Ibid., pp. 35–36.

42. LX, 7.423.

43. The arrows in question, lit. *shen shi* (divine arrows or the god's arrows) are guessed by most to be Cupid's. But no one thinks this is a love poem. Zhou Zhenfu implies that Cupid's arrows inspired in Lu Xun patriotic love, and thence "revoutionary thought." See Zhou Zhenfu, ed., *Lu Xun shi ge zhu* (Hangzhou 1980), pp. 24–39. Making Cupid's role somewhat more credible, an annonymous commentator insists that Lu Xun was aware of a refined version of the Cupid legend: Cupid shot arrows both of love and hate. Lu Xun, feeling both, had therefore "surpassed the limitations of the 'universal love' thinking of the ordinary advocates of democracy of the time." Be that as it may, if the arrow line means, "My heart cannot escape the arrows of love and hate," it could indeed describe Lu Xun's lifelong feelings toward China, whatever his specific loves and hates in 1903. See *Lu Xun shi jian xuanji* (Hong Kong, 1967), pp. 2–6. The third line is equally obscure, but most think it means that Chinese should not pin their hopes on the Guangxu emperor.

44. The question of Lu Xun's possible membership in one or more revolutionary parties is fuzzy in the extreme. See below, pp. 25–28.

45. LX, 1.416.

46. In August 1933, in an essay on essays, Lu Xun said China needed essays that were "daggers and spears," because Chinese Writers should all "struggle and fight" (figuratively). Chinese critics, ever since, have called his own essays "daggers and spears." See LX, 4.575 and 6.514. In January 1933 Lu Xun joined the Zhongguo Minquan Baozhang Tongmeng (The Chinese civil rights defence league), which had been founded by Song Qingling, Cai Yuanpei, Yang Quan, and others in December 1932. Lu Xun was elected to a nine-member executive committee of the Shanghai branch and regularly attended meetings. On June 18, however, the Guomindang assassinated Yang Quan. It was rumored that Lu Xun too was on the Guomindang's hit list, even that he was to be shot at Yang Quan's funeral. But Lu Xun went anyway. See Lu Xun Museum Lu Xun Research Office, ed., *Lu Xun nianpu* (Beijing, 1984), vol. 3, pp. 375–376, 424–425. In a letter dated the day of the funeral, June 20, 1933, Lu Xun said, "I hear the plan is to kill over ten other people. I cannot go about in the open" (LX, 12.188). But his diary suggests he already had (LX, 13.85). Lu Xun had predicted in February 1933, in a letter to a friend, that the Civil Rights Defense League would "probably not be long lived." After the assassination of Yang Quan, defenseless, it folded. See LX, 12.150–151.

3. To Change Men's Minds

1. A mixing of two accounts. See LX, 1.416 and 2.306.

2. *Alistair Cooke's America* (New York, 1973), pp. 312–313.

3. LX, 2.306.

4. Ibid., 1.417.

5. Zheng Yi, *Lu Xun sixiang fazhan lun gao* (Chengdu, 1981), p. 55.

6. LX, 1.417.

7. Ibid., 1.416.

8. Ibid., 2.306–307.

9. Zhang Binglin, who had cut off his queue as early as 1900, had been one of the leading revolutionary publicists in Shanghai working with Cai Yuanpei, Wu Zhihui, Zhang Shizhao, Zou Rong and others, through such organizations as the Zhongguo Jiaoyu Hui (The Chinese educational association), the Aiguo Xueshe, (The patriotic school), and the famous, if short lived, *Su bao* (Jiangsu journal). When the Qing government moved to close the *Su bao* in 1903, Zhang Binglin and Zou Rong were arrested and imprisoned. Zou Rong died in prison in 1905. Zhang Binglin was released in 1906, went to Tokyo, and became editor of Sun Yatsen's *Min bao* (The people's journal). See Mary Backus Rankin, *Early Chinese Revolutionaries*, (Cambridge, 1971), pp. 48–95. Lu Xun said in the last year of his life that he had studied with Zhang Binglin "not because he was a scholar, but because he was a scholarly revolutionary," but he confessed that what he had studied with Zhang Binglin was not revolution but philology. See LX, 6.546. Lin Fei and Liu Zaifu confirm this. See Lin, Liu, pp. 73–75. For the "oddity" of Zhang Binglin's thought, see CCD, pp. 413–420.

10. *Jingshen jie zhi zhanshi*, as fuzzy a phrase in Chinese as in English. See LX, 1.100 or Wang Shiqing, *Wu pian*, p. 183.

11. Lin, Liu, p. 64.

12. Zheng Yi, p. 55.

13. Wang Shiqing, *Lu Xun zhuan* (Beijing, 1981), p. 51.

14. Wang Shiqing, *Wu pian*, p. 262.

15. Ibid.

16. Ibid., p. 263.

17. Wang Shiqing, *Lu Xun chuangzuo daolu chu tan* (Beijing, 1981), pp. 9–10.

18. See above, p. 3.

19. John Hersey, *The Call* (New York, 1985), p. 84.

20. I am convinced that I read somewhere that this was a slogan used at the World's Student Christian Federation meeting held in Beijing in April 1922. I cannot, I confess, locate the phrase, but for a possible echo, see Hersey, pp. 351–352. For documentable evidence of similarly militant missionary rhetoric, consider Griffith

John's proclamation of the China missionary task in 1877: "We are here . . . to do battle with the powers of darkness, to save men from sin, and conquer China." A. R. Kepler, preaching a "social approach" to the "missionary enterprise," could nonetheless sound even more militant, or more modernly militant, in 1920: "Let us have good equiptment, mass our efforts, and then 'go over the top' and hit the enemy's lines hard." See Jessie G. Lutz, ed., *Christian Missions in China—Evangelists of What?* (Boston, 1965), pp. 11, 18, 21.

21. Hersey, p. 114.

22. Ibid., p. 135.

23. EE, p. 82.

24. Peter A. Kropotkin, *Mutual Aid* (London, 1919), p. 13.

25. O, p. 62.

26. Kropotkin, *Mutual Aid*, pp. 53–54 (phrases not quoted in consecutive order).

27. A most mysterious word, encompassing as it does every thing from the, one assumes, unconscious competition of genes to the, one assumes, conscious struggles of geniuses.

28. Kropotkin, *Mutual Aid*, p. 49.

29. *Meng Zi yi zhu* (Hong Kong, n.d.), 3.2, p. 62.

30. *Mao Zhuxi yulu* (Beijing, 1967), p. 11.

31. Ni Moyan, *Lu Xun geming huodong kao shu* (Shanghai, 1984), pp. 59–61.

32. Ibid., pp. 61, 64–65.

33. Ibid., p. 61.

34. Ibid., p. 63.

35. Ibid., pp. 65–66.

36. Ibid., p. 62.

37. *Huiyi Lu Xun ziliao jilu* (Shanghai, 1980), p. 50.

38. Ibid., p. 50.

39. See Rankin, pp. 183–185.

40. LX, 3.446.

41. *Mao xuan*, 3.906.

42. William A. Lyell, Lu Xun's Visions of Reality (Berkeley, 1976), p. 83.

•

43. Ni, p. 58.

44. Wang Shiqing, *Zhuan*, p. 54. The phrase that Wang Shiqing borrows from Lu Xun to describe the kind of warfare Lu Xun advocated is *"renxing de zhanzheng."* *"Renxing"* means "pliable but unbreakable." Lu Xun himself defined it with the classical idiom (from Xun Zi), *"qie er bu she"* (to carve away without quitting). Lu Xun seemed to mean that good fighters should not risk all in open warfare. They should not sacrifice themselves to no avail. They should "give," when necessary, but never give up. See LX, 1.164 and 11.46.

45. Li Zehou, *Zhongguo jindai sixiang shi lun* (Beijing, 1979), p. 443.

4. Deaf Ears

1. LX, 1.417.

2. Wang Shoushang, *Wang you*, p. 21.

3. Lin, Liu, p. 65.

4. *Henan* was founded in Tokyo in December 1907 by the overseas Chinese students Cheng Ke, Sun Zhudan and others. According to Lu Xun's brother, Zhou Zuoren, the editor-in-chief was Liu Shipei, republican revolutionary, anarchist, and later (per force?) turncoat (see Rankin, pp. 293, n. 115). Sun Zhudan, a friend of Zhou Zuoren from their Nanjing student days, asked both Zhou brothers to write for *Henan*, and they did. See *Huiyi Lu Xun ziliao jilu* (Shanghai, 1980), p. 49, and LX, 1.5 n. 2.

5. See LX, 4.11.

6. Ibid., 8.25.

7. See CCD, Chapter Six.

8. In February 1912, Lu Xun was invited by Cai Yuanpei, the new republics's first Minister of Education, to take a position in the new Ministry of Education. In May the Ministry moved to Beijing, and in June and July Lu Xun gave a series of four lectures, collectively entitled, "A Brief Discussion of the Fine Arts," in support of Cai Yuanpei's promotion of art education (Cai Yuanpei himself having suggested, rather grandiosely, "a substitution of aesthetics for religion"). Lu Xun's lectures, lost, alas, were not wildly succesful. At the first one, Lu Xun said, "thirty people listened, but five or six left in the middle." Twenty attended the second. No one came to the third, perhaps because Cai Yuanpei had just resigned and his policies had been overturned. Determined to deliver his lectures nonetheless, Lu Xun rescheduled his third lecture and got an audience again of twenty. At his last lecture, "at first there was only one person, but by the end there were ten," a thousand percent increase, but probably not very encouraging. See Ma Tiji, *Lu Xun jiangyan kao* (Haerbin, 1981), pp. 1–3.

9. For his veiled praise of Sun Yatsen see "Zhanshi he cangying" (Warriors and flies), LX, 3.88. For praise of Hu Shi see LX, 4.13. For less respectful references to Hu Shi, see LX, 5.46, 6.9, 12.155, 13.283.

10. See Ibid., 7.448.

11. See Ibid., 6.589.

12. I have been told by Chinese scholars who must remain for the time being, alas, anonymous, that in 1981 papers on Lu Xun's "loneliness" were not officially well received. Indeed the topic was close to being a "forbidden area." Thereafter, however, many scholars made pointed reference to Lu Xun's loneliness, clearly with their own loneliness in mind, the loneliness of Chinese intellectuals.

13. See LX, 8.24, and 1.99–100.

14. Ibid., 8.26.

15. Ibid., 1.417–418.

16. Ibid., 1.3.

17. Ibid., 7.4.

18. Ibid., 6.545.

19. Ibid., 6.546.

20. Zhao Ruihong, *Lu Xun 'Moluo shi li shuo' zhushi, jin yi, jieshuo* (Tianjin, 1982), p. 3.

21. "Freedom on the Wallaby," Henry Lawson, words, on Gordon Bok's "Bay of Fundy" (Sharon, Connecticut 1975), side II, band 6.

22. LX, 1.8–17.

23. Ibid., 1.25–35.

24. Ibid., 1.44–57.

25. Ibid., 1.63–100.

26. Ibid., 8.23–34.

27. See *CCD*, pp. 129–132, 426–427.

28. LX, 1.46.

29. Ibid., 1.45.

30. Ibid., 1.56.

31. Ibid., 1.52.

32. Ibid., 1.100.

5. The Riddle of the Universe

 1. I still, growl, cannot discover who first awarded Huxley this epithet.

2. R, pp. xi–xii.

3. LX, 1.8.

4. See Lu Xun, *Fen* (Hong Kong, 1964), p. 219 and LX 1.17.

5. Lin, Liu, pp. 66–67. Further evidence that Lu Xun had read, one way or another, *The Riddle of the Universe* can be found in "Po e sheng lun," (LX, 8.28–29), where Lu Xun describes Haeckel's monist religion (Cf. R, p. 336).

6. See LX, 1.8–9 for Lu Xun's translation from *The Riddle of the Universe*. For his borrowing of an argument from *The History of Creation*, see below, pp. 76–77.

7. Loren Eiseley, *Darwin's Century* (New York, 1958), p. 346.

8. R, p. vi.

9. Ibid., pp. 380–381.

10. HC, II, 367.

11. O, p. 484.

12. Ibid., p. 488.

13. Ibid., p. 489.

14. HC, I, 367.

15. R, p. 5.

16. HC, II, 332.

17. Ibid., 321.

18. R, p. 6. See also HC, I, 368.

19. R, pp. 382–383.

20. EM, II, 458.

21. HC, I, 368.

22. Like Haeckel I repeat myself, but that *is* our frightening reflection.

23. R, p. ix.

24. Huxley, *Science*, p. 318.

25. EE, p. 78.

26. EM, II, 438.

27. R, p. 304.

28. Ibid., p. 336.

29. See M title page.

30. HC, II, 368.

31. EM, II, 456.

32. M, p. 36. Huxley's famous phrase, from page 71 of *Man's Place in Nature*, was simply (but much more powerfully) "the question of questions."

33. R, p. 84.

34. HC, II, 278.

35. R, p. 211.

36. M, p. 3.

37. M, pp. 15–16.

38. R, p. 373.

39. R, pp. 228–229.

40. EM, II, 457.

41. M, p. 19.

42. *Lao Zi dao de jing*, 42, in *Wang Bi ji jiao shi* (Beijing, 1980), vol. I, p. 117. Lao Zi, better known as Lao Tzu, was (if he really existed) the first great Daoist (Taoist) philosopher.

43. HC, I, 35.

44. Ibid.

45. Ibid., p. 36.

46. Ibid.

47. R, p. 20.

48. M, pp. 18–19.

49. R, p. 288.

50. Quoted in R, p. 291.

51. R, p. 261.

52. HC, I, 35.

53. R, pp. 273–274.

54. Ibid., p. 14.

55. EE, p. 11. See above, p. 42.

56. R, p. 109.

57. EM, II, 454.

58. Ibid., p. 455.

59. R, p. 274.

60. Ibid., p. 381.

61. Ibid., p. 16.

62. Ibid., pp. 130–131.

63. Quoted in R, p. 383.

64. M, pp. 51–52.

65. Ibid., p. 50.

66. Ibid., p. 51.

67. Quoted in M, p. 51, from *Hamlet*, V, I, 3–6.

68. R, p. 208.

69. Ibid., p. 336.

70. Ibid., p. 382.

71. M, p. 86.

72. Ibid., p. 85.

73. O, p. 490.

74. M, p. 16. See above, p. 85.

75. R, p. 337.

76. From Alfred Tennyson, "In Memoriam," 56.

77. M, pp. 73–74.

78. R, p. 345.

79. EE, p. 53.

80. Ibid., p. 71.

81. Ibid., p. 52.

82. Ibid., p. 44.

83. M, pp. 85–86.

84. R, p. 309.

85. Ibid., p. 317.

86. Ibid., p. 351.

87. Ibid., p. 338.

88. Ibid., p. 353.

89. Ibid., p. 350.

90. Ibid., p. 338.

91. Ibid., p. 350.

92. See Richard Dawkins, *The Selfish Gene* (New York, 1978).

93. R, pp. 353–354.

94. Ibid., p. 351.

95. See O, 209: "If we suppose any habitual action to become inherited—and I think it can be shown that this does sometimes happen—then the resemblance between what was originally a habit and an instinct becomes so close as not to be distinguished."

96. HC, II, 364.

97. Ibid., p. 368.

98. Ibid., pp. 368–369.

99. Walt Kelly, *Pogo* (New York, 1951), p. 45.

100. HC, II, 369.

101. HC, I, 282.

102. James 4.11, Kurt Aland et al., eds., *The Greek New Testament* (New York, 1968), p. 787.

103. O, p. 340.

104. Ibid., p. 83.

105. Ibid., p. 489.

106. Ibid., p. 5.

107. As Lao Zi knew: "Heaven and earthare are not humane. They treat the ten thousand things as straw dogs." See *Lao Zi*, 5, in *Wang Bi*, I, 13.

108. O, pp. 79, 489.

109. Ibid., p. 80.

110. Ibid., p. 84.

111. Ibid., p. 471.

112. Ibid., p. 82.

113. Ibid., p. 102.

114. Ibid., p. 111.

115. Darwin, Voyage, p. 381. See also Jonathan Weiner's marvelous book, *The Beak of the Finch* (New York, 1994) for new evidence on "the waxing and waning" of beaks.

116. O, p. 205.

117. Francis Darwin, *The Life and Letters of Charles Darwin* (New York, 1889), vol. II, p. 67.

118. O, p. 490.

119. Herbert Spencer, *Universal Progress* (New York, 1864), p. 3.

120. O, p. 203.

121. O, p. 244.

122. Michael Flanders and Donald Swann, "Dead Ducks," *The Bestiary of Flanders and Swann*, side 2, #1.

123. R, pp. 273–274.

124. HC, I, 282.

125. Eiseley, p. 52.

126. HC, II, 367.

127. R, pp. 11–12, 14–15.

128. Ibid., p. 14.

129. EM, I, 6.

130. HC, I, 277.

131. Ibid., p. 309.

132. EM, I, 6. See also Stephen Jay Gould, *Ontogeny and Phylogeny* (Cambridge, 1977), p. 76. Gould says that "the law of recapitulation was 'discovered' many times in the decade following 1859." But Haeckel showed no doubt about its "true" discoverer.

133. Gould, *Ontogeny*, pp. 115–166.

134. EM, I, 3.

135. HC, I, 311.

136. EM, I, 6–7.

137. Ibid., pp. 7–8.

138. Ibid., p. 72.

139. See EM, I, 137. Haeckel did realize that "the eggs of Man, of the ape, of the dog, etc." must contain differences, although "even under the highest magnifying power of the best microscope" one could not see them. It was as if he anticipated some DNA like coding device. (He said indeed that the cause of our "delicate individual differences . . . must be sought only in the molecular structure" of "the ripe mammalian egg"). He was willing to admit that individuality was somehow present in the egg: "Even of human eggs, each differs from the other." But still one senses that he thought that the most important distinction in each egg was some sort of message that told it how far up the evolutionary tree it should climb before "getting off."

140. See EM, I, 114–115, or HC, I, 310.

141. Gould, *Ontogeny*, p. 3. Darwin himself pointed the way. See his explanation of his own maxim, that "community in embryonic structure reveals community of descent," in O, pp. 449–450.

142. HC, I, 310.

143. Ibid., p. 311.

144. See EM, II, plate XV opposite p. 188.

145. HC, II, 369.

146. Ibid., I, 174.

147. Ibid., II, 305–306.

148. Ibid., p. 325.

149. Ibid., pp. 327–328.

150. Ibid., pp. 294, 307, 308.

151. Ibid., p. 307.

152. Ibid., p. 310.

153. Ibid., p. 313.

154. Ibid., p. 307.

155. Ibid., p. 326.

156. EM, II, plate XIV opposite p. 180.

157. Ibid., p. 180–181.

158. HC, II, 309–310.

159. Ibid., p. 363.

160. Ibid., pp. 365–366.

161. See above, p. 39.

162. HC, II, 314.

163. And yet even today many Chinese believe that Caucasians exude, as an unfortunate racial trait, a *huchou* (fox stench)—why else their obsession with deoderants?

164. Ibid., p. 321, for the species' "history," and pp. 316–317. Except for the word "Mongol," my italics.

165. Ibid., p. 319.

166. Ibid., p. 321.

167. Ibid., p. 322.

168. Ibid.

169. Ibid., p. 323.

170. Ibid., p. 332.

171. Ibid.

172. Ibid., p. 324.

173. Charles Darwin, *The Descent of Man* (New York, n.d.), p. 53. In context, this quoted passage is particularly embarrassing for those of us who like Darwin,

because Darwin seems to call imperialism good. Moreover, in a line that would make Leninists lear, he seems to praise capital as the power behind imperialism. Elsewhere (see *Descent*, pp. 146, 178) he seems to predict the extermination of "savage races" without advocating it. But how does one explain this:

> The inheritance of property by itself is far from an evil; for without the accumulation of capital the arts could not progess; and it is chiefly through their power that the civilized races have extended, and are everywhere extending, their range, so as to take the place of the lower races.

6. The History of Mankind

1. Benjamin Schwartz, *In Search of Wealth and Power: Yen Fu and the West* (Cambridge, 1964), pp. 100–101.

2. See CCD, pp. 158–175.

3. See ibid. Part Three.

4. LX, 1.8.

5. Ibid.

6. Ibid., 1.9–10.

7. Carolus Linnaeus, *Systema Naturae* (London, 1956), pp. 20–21.

8. Wang Shiqing, *Wu pian*, p. 251.

9. LX, 1.12.

10. Ibid., 1.15—but on page 14 he got it right.

11. Ibid., 1.15.

12. See above, page 57.

13. R, p. 15.

14. LX, 1.8.

15. See above, p. 56.

16. Huxley, *Man's Place*, pp. 130–131.

17. Darwin, *Descent*, p. 707.

18. See Yan Fu, *Tianyan lun* (Taibei, 1967), Part 1, p. 48 and Part 2, p. 44.

19. A phrase from the national anthem of the People's Republic of China, "*Zhongguo renmin dao liao zui weixian di shihou*" (The Chinese people have reached

their most dangerous hour). The hour in question was that of the Japanese invasion of China that began, in Asia, World War II. The present Chinese national anthem was originally a patriotic song from that horrible period.

20. As do the Japanese. I am still not sure who first used the term.

21. "If names are not correct, discourse is difficult." *Lunyu yi zhu*, ed. Yang Bojun (Beijing, 1965), 13.3, p. 140.

22. LXYJ, 3.122.

23. See above, p. xii.

24. LX, 1.8, 12–15.

25. HC, I, 16–17.

26. O, p. 6.

27. HC, I, 100.

28. LX, 1.9.

29. Ibid., 1.13.

30. Ibid., 1.67.

31. See, for example, Liu Zaifu, Jin Qiupeng, Wang Zichun, *Lu Xun he ziran kexue* (Beijing, 1979), pp. 107–108.

32. LX, 1.66.

33. See CCD, pp. 53–56.

34. LX, 1.67–68.

35. See CCD, pp. 56, 92–93.

36. See above, p. 16.

37. LX, 1.67.

38. LX, 1.28.

39. *Mao yulu*, p. 168.

40. Hou Yi was the mythical, one assumes, archer who, in ancient times, shot nine of the ten suns out of the sky.

41. Yan Fu, Part 1, p. 48.

42. Ibid., II, 44.

43. LX, 1.14.

44. Ibid., 8.32.

45. Ibid., 7.230.

46. Ibid., 4.5. Yan Fu had written in *Tianyan lun,* in his first comments to Huxley's "Prolegomena," "What we can know is that the way of the world must progress, and that the future will surpass the present." See Yan Fu, Part 1, p. 48.

47. *Mao xuan,* 3.932.

48. LX, 1.322.

49. Ibid., 1.17.

50. Ibid., 1.14.

51. Ibid., 1.13.

52. Ibid., 1.8.

53. Ibid., 1.12.

54. Ibid., 1.8.

55. *Mao yulu,* p. 75.

56. Wang Shiqing, *Wu pian,* pp. 253, 21.

57. *Meng Zi,* 1.7, p. 15.

58. LX, 4.575.

59. See below, pp. 262–264.

60. Dr. Seuss, *Horton Hatches the Egg* (New York, 1940), n.p.

7. On Human Nature

1. LX, 8.32.

2. Ibid., 8.31–32.

3. *Mao yulu,* pp. 67–68.

4. LX, 8.31.

5. For this patchwork of quotations see EE, 52, 85, 43–44, 27, 85, 44. Huxley's "ape and tiger" were actually Tennyson's, of course. See above, Chapter I, n. 23. One can gain more than an inkling of what Huxley meant by his proto-Freudian "ape and tiger promptings" from a marvelously paradoxical paragraph from an essay he wrote in 1892, a year before "the ape and tiger" made their appearance in "Evo-

lution and Ethics," a paragraph that reveals as well how "biblical" Huxley remained not only in his ethics but in his philosophy, despite the power of his attacks on the infalliblity of the Bible:

> That the doctrine of evolution implies a former state of innocence of mankind is quite true; but, as I have remarked, it is the innocence of the ape and of the tiger, whose acts, however they may run counter to the principles of morality, it would be absurd to blame. The lust of the one and the ferocity of the other are as clear evidences of design, as any other features that can be named [See Huxley, *Science*, p. 52].

Surely by "design" he meant "natural selection," but even so we have a very strange "fall of man," through a "loss of innocence," through knowledge of "the reality" of good and evil (see EE, P. 52). But from what tree of knowledge did Huxley pluck the double standard by which he measured in "the innocence of the ape and tiger" *our* "original *sin?*"

6. See Cheng Ma, "Lu Xun liu Ri shiqi de qimengzhuyi sixiang jiegou," in LXYJ, 8.76. I have not chased down the first Japanese translation of Also *Sprach Zarathustra*, but even the word Lu Xun used for "worm" (see subsequent text) suggests Japanese help. "*Quchong*" is very obscure Chinese. Read "*ujimushi*" it is apparently much less obscure Japanese. When or if Lu Xun read Zarathustra in the original, I have been unable to determine. In the quotation from Nietzsche that opens "Moluo shi li shuo" (see above, p. 80), Zhao Ruihong notes discrepancies between Lu Xun's classical Chinese translation and the German original, but whether this represents Lu Xun's trouble with the German original, or a Japanese translator's, or the dangers of double translation, is uncertain (see Zhao Ruihong, *Lu Xun "Moluo shi li shuo"* [Tianjin, 1982], p. 18). In 1918 Lu Xun translated into classical Chinese the first three sections of "Zarathustra's Prologue," but did not publish his translation. He did publish a vernacular translation of the whole prologue in 1920, and in a postscript provided some key German terms, so by then he had at least consulted the original (see Lin, Liu, p. 124, and LX, 10.439).

In any case, Lu Xun never "recovered" from his encounter with Nietzsche. He first mentioned him, and quoted him, in 1907. He would go on mentioning him, quoting him, and imitating him for the rest of his life. Lu Xun *liked* Nietzsche. He read Nietzsche with an excitement that clearly effected his own writing.

That fact was for years an embarrassment to upholders of the faith in the People's Republic. Because Nietzsche was assumed to be an enemy of "the masses," and because nasty people liked Nietzsche, it was assumed that Nietzsche himself was nasty, a proto-Nazi. In 1982 Zhao Ruihong said, "Research into the relationship between Lu Xun and Nietzsche—intellectual and artistic—has actually become a 'forbidden area'"—off limits (Zhao Ruihong, p. 282). But Mr. Zhao himself helped open up that area. Scholars might still ask of Lu Xun and Nietzsche what we are now, I confess, asking of Lu Xun and Haeckel: "What made Lu Xun— a patriotic youth from 'a country the victim of aggression'—feel well disposed toward Nietzsche's philosophy, that possessed such poison?" (Jin Hongda, *Lu Xun wenhua sixiang tansuo* [Beijing, 1986], p. 63). But scholars are now willing to argue that "Lu Xun's passion for Nietzsche never waned throughout his life" (Min Kang-sheng, "'Guoke' and *Also Sprach Zarathustra*," a paper delivered at the Conference on Lu Xun and Chinese and Foreign Culture, held in Beijing from October 19–24, 1986, in commemoration of the fiftieth anniversary of Lu Xun's death, p. 1). Others have even said that Lu Xun "completely accepted Nietzsche's thought," that "some of Lu Xun's essays are virtually parts of *Thus Spoke Zarathustra* rewritten," and that consequently "it is indeed not without reason that Lu Xun has been called China's Nietzsche" (See Zhang Zhaoyi, "Jinhua lun yu chaoren de maodun," in LXYJZL, 17.327).

Granted it is difficult to separate influence from resonance, especially as resonance can be in the ear of the listener—Zhang Zhuo has said that Lu Xun liked Nietzsche because he misunderstood him (see Zhang Zhuo, *Lu Xun zhexue sixiang yanjiu* [Hubei, no city, 1981], pp. 58–62)—but whether Lu Xun understood Nietzsche or misunderstood him (we shall have to see), he liked what he heard, and he echoed it. We can hear Nietzsche in Lu Xun's works, over and over again. We can hear his language. We can hear his *Darwinian* language—although neither Lu Xun nor Nietzsche was a good Darwinian.

7. LX, 8.32.

8. Friedrich Nietzsche, *Thus Spoke Zarathustra* (Baltimore, 1968), p. 42.

9. See LX, 6.238.

10. EE, p. 44.

11. LX, 8. 32. The "great unity" was the *da tong*, the ancient utopian vision appropriated by both Kang Yuwei and Sun Yatsen. See CCD, pp. 18–21, 31–42.

12. A mixture of Haeckel's language and Lu Xun's. See above, p. 65.

13. See above, p. 65.

14. Veblen cited in Richard Morris, *Evolution and Human Nature* (New York, 1984), p. 53. For Haeckel's influence on Lombroso, see Gould, *Ontogeny*, pp. 120–126.

15. Quoted in Richard Morris, p. 109.

16. Ibid., pp. 109–110.

17. Konrad Lorenz, *On Aggression* (New York, 1963), p. 30.

18. Ibid., p. x.

19. Sigmund Freud, *Civilization and Its Discontents* (New York, 1962), p. 69. See also p. 8.

20. Lorenz, pp. 30–31.

21. EE, p. 27.

22. *Xun Zi jian shi*, ed. Liang Qixiong (Taibei, 1969), Chapter 23, p. 338.

23. EE, pp. 51, 85.

24. Freud, pp. 58–59.

25. Lorenz, p. 241.

26. See Freud, p. 58. Plautus's line (*Asinaria*, l. 495) was "*Lupus est homo homini, non homo, quom qualis sit non novit*" (A wolf man is to man, not a man, when he knows not of what sort he is).

27. Lorenz, p. 129.

28. Ibid., pp. 241–243.

29. Ibid., p. 269.

30. Ibid., p. 273.

31. *Luo chong*. See *Ci hai* (Shanghai, 1979), vol. 1, p. 576. All featherless, furless, scaleless, and shelless creatures were "naked beasts," but human beings were best of the lot.

32. See Desmond Morris, *The Naked Ape* (New York, 1984), Chapter 5 and pp. 196, 152.

33. Philip Appleton, ed., *Darwin* (New York, 1979), p. 439.

34. Peter Kropotkin, *Memoirs of a Revolutionist* (New York, 1970), p. 499.

35. Kropotkin, *Mutual Aid*, p. 64.

36. Freud, p. 69.

37. LX, 1.432.

38. Ibid., 11.31.

39. *Mao xuan*, 3.827.

40. D. W. Fokkema, *Literary Doctrine in China and Soviet Influence* (The Hague, 1965), pp. 236–243.

41. LX, 1.432.

42. Ibid., 1.358.

43. Nietzsche, pp. 103, 43.

44. Ibid., p. 124.

45. Ibid., pp. 112, 41, 100.

46. LX, 8.33 (my italics).

47. Ibid., 1. 422, 424–425. It will be evident in the pages that follow that Lu Xun used his phrase, "eat people," in "a large and metaphorical sense"—as Darwin said of his famous phrase, "Struggle for Existence" (See O, p. 62). Nevertheless, Lu Xun cited some instances of cannibalism in China that were real enough. The revolting revelations in 1992 by Zheng Yi and Nicholas Kristof of real cannabilism—for grotesque political reasons—in Guangxi during the Cultural Revolution would have made Lu Xun doubly sick, but would probably not have surprised him. See Nicholas D. Kristof and Sheryl WuDunn, *China Awakes* (New York, 1994), pp 73–75.

48. Ibid., 1.526.

49. Ibid., 1.423–424.

50. See Ibid., 2.5 and 1.434.

51. The Revolution of 1911 succeeded in overthrowing the Qing Dynasty and ending the monarchy, but before a year was out, Yuan Shikai, the "Republic's" first president, had made a tragic farce out of republican government. In its democratic hopes, Sun Yatsen's revolution has, in his all too immortal dying words, "still not succeeded."

52. LX, 1.429.

53. Ibid., 1.325.

54. See Walt Kelly, Pogo: *We Have Met The Enemy And He Is Us*.

55. LX, 6.239.

56. John Stuart Mill, *On Liberty* (New York, 1947), p. 72.

57. LX, 1.430–431.

58. Nietzsche, p. 75.

59. LX, 1.430.

60. *Meng Zi*, 9.7, p. 225.

61. LX, 1.135.

62. Ibid., 1.50.

63. Ibid., 1.68.

64. Ibid., 1.52.

65. Ibid., 1.166.

66. Ibid., 1.52.

67. Ibid., 1.432.

68. Ibid., 1.422.

69. Ibid., 1.432.

70. Ibid., 1.419.

71. Ibid., 8.23.

72. Ibid., 1.100.

73. Ibid., 7.230. See above, p. 74.

74. Ibid., 4.5. Again see above, p. 74.

75. Ibid., 1.423.

76. EE, p. 85.

77. *Meng Zi*, 8.19, p. 191, 8.12, p. 189.

78. See above, p. 80.

79. *Lun heng zhushi*, ed. Peking University History Department *Lun heng* Annotation Committee (Beijing, 1979), p. 201.

80. LX, 1.272.

81. Quoted in Hofstadter, pp. 47–48.

82. Stephen Jay Gould, *The Flamingo's Smile* (New York, 1985), p. 241.

83. Stephen Jay Gould, *The Panda's Thumb* (New York, 1982), pp. 182, 212–213.

84. Ibid., pp. 182, 213.

85. O, p. 194.

86. Gould, *Panda*, p. 182.

87. Ibid., p. 212.

88. Ibid., p. 213.

89. Stephen Jay Gould, *Ever Since Darwin* (New York, 1979), p. 251.

90. Wilson, *Sociobiology*, p. 4.

91. Darwin's usual word for physiological traits. See O, pp. 154, 413, etc.

92. *Pimao* (literally skin and hair) is a common Chinese metaphor for superficial things—a good word to describe racial differences.

93. Quoted in Gould, *Ever Since*, p. 247. Mill's full sentence, which Gould thinks should be adopted as "the motto of the opposition" (to advocates of "bilogical determinism in the study of human intelligence"), went as follows:

> Of all the vulgar modes of escaping from the consideration of the effect of social and moral influences upon the human mind, the most vulgar is that of attributing the diversities of conduct and character to inherent natural differences.

Lu Xun did *not* seek any such vulgar mode of escape. Only when he was most depressed did he echo the language of biological determinism. Most of his writing was specifically aimed at "social and moral influences upon the [Chinese] mind."

94. *"Ke ji fu li wei ren"* (to conquer oneself and return to the rites is [the way to become] humane). See *Lunyu*, 12.1, p. 130.

95. Ibid., 17.2, p. 188.

96. Ibid., 17.3, p. 188.

97. Ibid., 4.6, p. 39.

98. *Xun Zi*, Chapter 23, p. 336. Yao and Yu were two of the legendary three sage kings.

99. Ibid., p. 337.

100. *Meng Zi*, 12.2, p. 276. Shun was the second of the three sage kings.

101. Ibid., 1.7, p. 15.

102. *Lunyu*, 7.30, p. 80.

103. Asa Gray, *Darwiniana* (Cambridge, 1963), p. 140.

104. G. F. Handel, *The Messiah* (New York, 1912), pp. 218–219.

105. O, pp. 88–89.

106. Ibid., p. 208.

107. Charles Darwin, *The Expression of the Emotions in Man and Animals* (Chicago, 1965), p. 354.

108. Ibid., p. 350.

109. Ibid., pp. 10–11.

110. Ibid., p. 352.

111. Ibid., p. 356.

112. Ibid., p. 12 (my italics).

113. O, pp. 208–209.

114. From Lorenz' "Preface" to Darwin, *Expression*. See pp. xi, xii.

115. Edward O. Wilson, *On Human Nature* (New York, 1979), p. xiii.

116. Ibid., p. 14.

117. Ibid., p. 5.

118. Ibid., p. 6.

119. See above, p. 91.

120. Wilson, *On Human Nature*, p. 2.

121. Ibid., p. 1.

122. Dawkins, p. 49.

123. Wilson, *On Human Nature*, p. 73.

124. O, p. 207.

125. "I will," *not* "I fly."

126. Gould, *Ever Since*, p. 252.

127. Ibid., p. 251.

128. Ibid., p. 254.

129. Ibid., pp. 257–258.

130. Ibid.

131. Ibid., p. 252.

132. Ibid., p. 259.

133. Ibid., p. 257.

134. Alexis de Tocqueville, *Democracy in America* (New York, 1945), II, 352.

135. EE, p. 80.

136. Gould, *Ever Since*, p. 266.

137. Ibid., p. 257.

138. Ibid., p. 266.

139. Ibid.

140. Ibid., p. 237.

141. From an unknown child pedant's version of "Row, Row, Row, Your Boat," sung at Lake Okiboji circa 1950:

> Propel, propel, propel your craft
> Placidly down the liquid solution.
> Ecstatically, ecstatically, ecstatically, ecstatically,
> Existence is but an illusion.

142. Stephen Jay Gould, *Hen's Teeth and Horses Toes* (New York, 1983), pp. 262–263.

143. *Meng Zi*, 11.1, p. 253.

144. Ibid., 11.2, p. 254.

145. B. F. Skinner, *Walden Two* (New York, 1976), p. vi.

146. Gould, *Ever Since*, p. 259.

147. Skinner, p. 242.

148. Gould, *Ever Since*, p. 257. My italics, except for *are*.

149. See *Lunyu*, 20.1, p. 215, for *ren ren*. For *renzhe, ren ye*, see *Zhongyong* 20, in *Si shu xin jie*, ed. Zhang Shoubai (Tainan 1961), p. 38. In slightly different form, but identical in meaning, *ren ye zhe ren ye*, the line also appears in *Meng Zi*, 14.16, p. 329. Its beautiful simplicity is made possible by a powerful pair of homonyms— which, in Chinese, the eye, if not the ear, can distinguish. The first *"ren"* is "humanity" or "humaneness" or "benevolence" or "love," and the second is "man" (in the generic sense, of course). So the sentence means "Humanity means humanity," or "To be humane is to be human."

8. Evolution and Ethics Again

1. *The Songs of Michael Flanders and Donald Swan* (New York, 1977), p. 64.

2. *Xin min congbao*, I, 2:34, quoted in *CCD*, p. 311.

3. Liang Qichao, *Yin bing shi wenji* (Taibei, 1960), III, 6:39, quoted in CCD, p. 315.

4. *Qing yi bao*, ed. Liang Qichao, Feng Jingru et al. (Taibei, 1967), XII, 6564. Quoted in *CCD*, p. 316.

5. EE, p. 52. See above, Chapter VII, n. 5.

6. See again EE, pp. 82, 83, 44.

7. See again Ibid., pp. 11, 59, 81.

8. Jean-Paul Sartre, *Existentialism and Human Emotions* (New York, 1957), p. 23.

9. "The Russian Tolstoy's doctrine of pacifism will not work," said Lu Xun in 1927. See LX, 8.190. And in 1928 he said, "Struggle I do think correct. If people are oppressed, why should they not struggle?" See LX, 4.83.

10. LX, 8.34.

11. Ibid., 1.4.30–431.

12. Ibid., 8.32.

13. Ibid., 8.34.

14. Ibid., 8.33.

15. Zhou Enlai was the more reassuring, but the phrase was the Chairman's.

16. See *Lunyu*, 4.15, p. 41, and 15.24, p. 173.

17. *Mo Zi jian gu*, ed. Sun Yirang (Taibei, 1970), Chapter 16, p. 78. Quoted in CCD, p. 253.

18. Dawkins, p. 78.

19. Wilson, *Sociobiology*, p. 551.

20. Dawkins, p. 179.

21. See Wilson, *Sociobiology*, p. 120 and Dawkins, p. 179.

22. Dawkins, p. 90.

23. LX, 8.33.

24. Ibid., 8.32.

25. EE, pp. 79–80 (my italics).

26. Wilson, *Sociobiology*, p. 287.

27. LX, 1.30. The modern Chinese word for intuition is *zhijue*, "direct sense." Lu Xun's word, which I have never seen elsewhere, was *shengjue*, a term that quite defies literal translation. It means something like "sagely sense," defined by Huxley, said Lu Xun, as "the discoverer of truth." See LX, 1.30. Wang Shiqing equates it with *linggan*, the modern word for "inspiration" (see Wang Shiqing, *Wu pian*, p. 87), but that is still something that gives one a "direct sense," an intuitive sense, of "truth"— or something.

28. EE, p. 71.

29. See above, Chapter 2, n. 38.

30. *Mao yulu,* p. 54.

31. Lu Xun, *Fen,* p. 285, n. 15.

32. Sartre, p. 45.

33. LX, 1.99, 1.68.

34. Ibid., 1.57.

35. Ibid., 1.29.

36. Gray, p. 133.

37. *Meng Zi,* 14.37, p. 341.

38. LX, 1.274.

39. Lu Xun, ed., *Ji Kang ji* (Hong Kong, 1967), p. 41.

40. Ibid.

41. LX, 3.563.

42. Ibid., 3.532.

43. Ibid., 4.560–561.

44. Ibid., 11.16.

45. Ibid., 3.501–517.

46. Ibid., 3.264.

47. *Lu Xun nianpu* (Hefei), p. 297.

48. See, for example, LX, 3.297.

49. *Lu Xun nianpu* (Hefei), p. 343, and Lloyd E. Eastman, *The Abortive Revolution* (Cambridge, 1975), p. 7.

50. *Xu Guangping yi Lu Xun,* Ma Tiji, ed. (Guangzhou [?], 1979), p. 639.

51. Ma Tiji, *Lu Xun jiangyan kao* (Haerbin, 1981), pp. 245–246.

52. Ibid, p. 247. For Wu Han, see Chapter I, n. 1.

53. LX, 3.514.

54. Ma Tiji, pp. 230–231.

55. In 1982 I had the great good fortune to be taken to visit the writer Xiao Jun, one of Lu Xun's "disciples" and to the end, as he proclaimed himself that day, "Lu Xun *pai,*" a "Lu Xunite." I asked him, hoping to save myself some work, what the most important studies of Lu Xun were that I should read. He replied, although

he had written some of them himself, "Don't read any of them. Just read Lu Xun." I did not take his advice, and I rather hope you will not, but—read Lu Xun.

56. I borrow, of course, Mao Zedong's metaphor—from his most famous utterance: "*Geming bushi qingke chifan*" (Revolution is no dinner party, or, more literally, revolution is not inviting guests to dinner). See *Mao yulu*, p. 11. Lu Xun seemed to accept matter-of-factly the roughness of "nature red in tooth and claw," without any expressed need for the consolation Darwin offered in his great whitewash of the bloodiness of the struggle for existence: "When we reflect on this struggle, we may console ourselves with the full belief, that the war of nature is not incessant, that no fear is felt, that death is generally prompt, and that the vigorous, the healthy, and the happy survive and multiply." See O, p. 79. The Chairman seemed to feel no need of consolation reflecting on the struggle either of evolution or revolution.

57. *Mao xuan*, 4.1365.

58. Ibid., 3.828.

59. He once wrote: "I suspect that among those who have suffered are young people who have bravely revolted having read and been effected by my articles. That gives me great pain. But that is because I am not a natural born revolutionary" (LX, 4.98). To Xu Guangping he once wrote: "As for social combat, I do not go off to battle. That is why I do not urge others to sacrifice themselves and so forth" (LX, 11.16).

60. LX, 4.83.

61. Ibid., 8.32.

62. Ibid., 1.272–273.

63. Ibid., 1.276.

64. *Mo Zi*, Chapter 39, p. 183.

65. If I may run together two of Richard Dawkins' phrases. Chapter Four of his *The Selfish Gene* is entitled, "The Gene Machine." See Dawkins, p. 49.

66. Numbers 31.15–18.

67. *Lunyu*, 12.19, p. 136.

68. Ibid., 13.11, p. 144.

69. LX, 1.274.

70. *The Confessions of St. Augustine*, eds. James Marshall Campbell and Martin R. P. McGuire (Englewood Cliffs, 1959), p. 159.

71. See Exodus 21.12, 22.19.

72. LX, 11.20.

73. Ibid., 1.276.

74. Ibid., 1.273.

75. *Mao yulu*, p. 75.

76. LX, 13.117.

77. Ibid., 1.526. See above, p. 86.

78. Ibid., 1.272.

79. See above, p. 96.

80. LX, 1.277.

81. Gould, *Ever Since*, p. 254.

82. LX, 1.358.

83. Spencer, *Social Statics*, p. 149.

84. Ibid., pp. 149, 146.

85. Quoted in Hofstadter, p. 41.

86. LX, 1.275.

87. Spencer, *Social Statics*, p. 152.

88. Ibid., pp. 151, 150.

89. LX, 7.398.

90. Ibid., 1.275.

91. Ibid., 13.250.

92. Spencer, *Social Statics*, p. 149.

93. *Lao Zi*, Chapter 5, pp. 13–14.

94. Fyodor Dostoyevsky, *Crime and Punishment* (London, 1988), p. 276.

95. LX, 1.274–175.

96. Ibid., p. 1.277.

97. See above, p. 215.

98. LX, 1.275.

99. Ibid., 1.216–217.

100. *Lu Xun zhi Xu Guangping shujian* (Shijiazhuang, 1979), p. 109.

101. LX. 3.423.

102. [*Lu Xun yulu*], pp. 5–6.

103. See above, pp. 111.

104. Gertrude Himmelfarb, *Victorian Minds* (New York, 1970), p. 324.

105. *Mao yulu*, p. 8.

106. Ibid., p. 10.

107. See Richard Herrnstein, "I. Q.," in *The Atlantic Monthly*, 228 (1971), pp. 43–64.

108. Lucian Pye, *China: An Introduction* (Boston, 1972), p. 254.

109. *The Chinese-English Dictionary*, English Department, Beijing Foreign Language Institute, ed. (Hong Kong, 1979), p. 111.

110. LX, 8.162.

111. Ibid., 3.454.

112. Ibid., 5.205.

113. Ibid., 4.282.

114. LXYJ, 3.230.

115. See above, p. 88.

116. For an excellent account of the rise and fall of the "Liang Xiao" group, which "flourished" from 1973 to 1976, see Yue Daiyun and Carolyn Wakeman, *To the Storm* (Berkeley, 1985), Chapter 14, pp. 322–348.

117. Zhang Yu'an, "Jiaru ta hai huozhe," in *Renmin ribao* (Beijing, October 20, 1980), p. 8.

118. LX, 3.423.

119. Ibid., 11.39.

120. Ibid., 11.43 (Xu Guangping's "three inch sword" allusion still eludes me), and Ibid., 11.45–46.

121. Ibid., 6.588–589.

122. *Mao yulu*, p. 11.

123. See above, Chapter 7, n. 149.

124. *Lunyu*, 12.22, p. 138. The "humanity" in Fan Chi's question means "humaneness." The "humanity" in the master's answer means "human beings."

125. *Meng Zi*, 7.2, p. 165.

126. Schwartz, p. 15.

127. Granted, neither Confucius nor Mencius was an absolute pacifist. See text below and notes 124 and 125.

128. As Bo Yi and Shu Qi said (or sang) of Wu Wang's overthrow of the Shang tyrant Zhou in Sima Qian's *Shi ji*. See *Shi ji hui zhu kaozheng (Shiki kaichu kosho)*, Takigawa Kametaro, ed. (n.p. n.d.), vol. 7, p. 10.

129. *Meng Zi*, 11.13, p. 272.

130. See Ibid., 2.6, p. 40 and 2.8, p. 42.

131. See above, p. 113. Confucius did, after all, once entreat the Duke of Lu to make war on Chen Heng, the rebel who had killed the Duke of Qi (see *Lunyu*, 14.21, p. 160).

132. *"Ruo rou qiang shi."* Put together at least by early Ming times from a sentence by the Tang poet Han Yu. See *Ci hai*, p. 2498.

133. LX, 1.80.

134. See above, p. 117.

135. Mark 4.23, *Greek New Testament*, p. 135.

136. Yet another formerly ubiquitous slogan, the *locus classicus* of which refuses to be found.

137. See the story of Xiang Yu, chief unsuccessful contender for the throne after the fall of Qin. On the night before his last battle "he heard from the Han army on all four sides the songs of Chu," his own state, so many of his people having gone over to the enemy. See *Shi ji*, 2.68.

138. See above, p. 87.

139. Galatians, 5.15, *Greek New Testament*, p. 660.

140. From Aeschylus, *The Eumenides*, in *Aeschyli septem quae supersunt tragoedias*, ed. Denys Page (Oxford, 1972), pp. 283–284.

141. LX, 6.636.

142. Ibid., 13.435.

9. The Evolution of Lu Xun

1. I do not know who first coined the terms "Demokelaxi Xiansheng" (Mr. Democracy) and "Saiyinsy Xiansheng" (Mr. Science), happily shortened to "De

Xiansheng" and "Sai Xiansheng," but they were widely used in the polemics that helped lead up to the May Fourth Movement of 1919. Most ironically, it seems now, Chen Duxiu, half a year before he declared himself a Marxist, used Mr. Science and Mr. Democracy to great advantage in defense of his famous magazine *Xin qingnian* (New Youth) in January 1919, in an article that could have been read to good effect on May 4, 1989 (See Chow Tse-tsung, *The May Fourth Movement* [Cambridge, 1960], p. 59). Mr. De and Mr. Sai were in Tiananmen Square again that day—and remained there until June 4.

2. LX, 1.56.

3. Ibid., 1.311.

4. Ibid., 8.89.

5. Ibid., 1.313.

6. Ibid., 5.479–480.

7. Liu, Jin, Wang, p. 15.

8. LX, 1.301.

9. Ibid., 6.591.

10. Ibid., 1.307.

11. Ibid., 7.253.

12. Ibid., 4.15.

13. Ibid., 1.306.

14. Ibid., 1.345.

15. Ibid., 1.314.

16. Ibid., 11.354.

17. Ibid., 1.325.

18. Ibid., 3.418.

19. Ibid., 1.359.

20. Ibid., 1.338–339.

21. *Mao yulu*, p. 10.

22. LX, 3.418.

23. Ibid., 4.240.

24. Ibid., 4.251.

25. Liu, Jin, Wang, p. 282.

26. See above, Chapter 7, n. 141.

27. See above, p. 102.

28. From Zhu Xi's "Guan xin shuo" (On viewing the mind). See *Zhongguo zhexue shi ziliao xuanji, Song, Yuan, Ming zhi bu*, Chinese Philosophy Historical Research Group of the Philosophy Institute of the Chinese Academy of Social Sciences, ed. (Beijing, 1982), p. 278.

29. LX, 3.45.

30. Ibid., 3.51, 3.45.

31. See above, Chapter 8, n. 130.

32. *Mao Zedong lun wenxue yu yishu* (Beijing, 1964), p. 101.

33. *Mao yulu*, p. 10.

34. *Mao xuan*, 5.42.

35. *Renmin ribao haiwai ban*, June 28, 1989, p. 1.

36. LX, 6.612.

37. [*Lu Xun yulu*], p. 22.

38. LX, 5.480.

39. [*Lu Xun yulu*], p. 2. "Marxismized is a literal rendition of the Chairman's grotesquery, Makesizhuyihua.

40. Ibid., p. 25.

41. Ibid., p. 26, or see LX, 6.589. I suppose one could translate the rather odd *"Mao Zedong xianshengmen"* a bit less literally, as "Mr. Mao Zedong et al."

42. [*Lu Xun yulu*], pp. 25, 27.

43. Ibid., p. 5.

44. *Mao yulu*, p. 11.

45. Suffice it to say that "critics" of party leaders and party policies have not all been either "anti-communists" or "counterrevolutionaries." Party leaders have for so long forced their critics into those categories, however, they have now indeed forced many, at least in their hearts, to rebel: *guan bi min fan.*

46. *Wenxue yundong shiliao xuan* (Shanghai, 1979), vol. 4, pp. 573–574. See also Merle Goldman, *Literary Descent in Communist China* (Cambridge, 1967), Chapter 2.

47. The most often repeated story is that Wang Shiwei was shot in 1947, when the Communists were driven out of Yan'an. See Simon Leys (Pierre Ryckmans), *Chinese Shadows* (New York, 1977), p. 127. Dai Qing, however, without clearly stating her source of imformation, maintains that Wang Shiwei was executed with a hatchet. See Dai Qing, *Wang Shiwei and "Wild Lilies"* (New York, 1994), pp. 3, 66. Mao Zedong confessed in a "Talk at an Enlarged Central Work Conference," on January 30, 1962, that Wang Shiwei had been "executed," but protested that "the security organs themselves made the decision to execute him; the decision did not come from the Centre." See Stuart Schram, ed., *Chairman Mao Talks to the People* (New York, 1974), p. 185. For a discussion of who might have given the order, see Dai Qing, pp. xix, 66–69.

47. LX, 4.575.

48. Ibid., 6.514. A good example of Lu Xun's "incendiary," or "inflamatory" language.

49. Pusey, *Wu Han*, p. 13 (translation slightly altered).

50. The article by Yao Wenyuan attacking Wu Han in November 1965 was the prelude to the Cultural Revolution (See Pusey, *Wu Han*, Chapter 7). Wu Han was "sent down to the countryside" near Beijing in March 1966. Then he was dragged here and there to be struggled almost every day for the better part of two years. And, finally, in March 1968, he was arrested and imprisoned. On October 11, 1969, he died mysteriously, but not too mysteriously, in prison. See Su Shuangbi and Wang Hongzhi, *Wu Han zhuan* (Beijing, 1984), pp. 324–333.

51. Kyna Rubin, "Keeper of the Flame: Wang Ruowang as Moral Critic of the State," in Merle Goldman, ed., with Timothy Cheek and Carol Lee Hamrin, *China's Intellectuals and the State: In Search of a New Relationship* (Cambridge, 1987), p. 241.

52. Ibid., p. 236.

53. Ibid., pp. 243–244.

54. *Wenxue yundong*, vol. 4, p. 594.

55. LX, 7.113. The word I here translate as "literature," so as to have one entity to contrast with "politics," in both Lu Xun's title and Wang Ruowang's, is *wenyi*, literally "literature and art."

56. Another slogan, formerly quoted ad nauseam, the origin of which I cannot discover.

57. LXYJ, 6.221.

58. Ibid., 3.230.

59. Ibid., 6.220.

60. Ibid., 1.4. See above p. xiii, note 18.

61. Ibid., 8.378.

62. Ibid., 3.242.

63. Yi Zhuxian, *Lu Xun sixiang yanjiu* (Wuhan, 1984), p. 21.

64. LXYJ, 3.245.

65. Ibid., 3.242.

66. LXYJZL, 7.39.

67. See Lu Xun's essay, "Ma sha yu peng sha" (Killing with curses and killing with praise), LX, 5.585. Scholars from all sides have played with these phrases. See for example Han Shaohua's attack on the Gang of four (LXYJ, 8.379) and Ni Moyan's attack (when the Gang held sway) on Zhou Yang and "swindlers like Liu Shaoqi" (*Xuexi Lu Xun geming dao di* [Shanghai, 1973], p. 147).

68. *Renmin ribao*, September 27, 1981, p. 4.

69. Yi, p. 21.

70. LXYJ, 1.5.

71. Quoted, for example, in *Lu Xun yanlun xuanji* (Beijing, 1976), vol. 3, p. 1. I think the Chairman made this remark in the middle of the Cultural Revolution, but I do not know exactly when or on what occasion.

72. LXYJ, 8.378.

73. See *Lu Xun zawen xuanjiang* (n.p., 1973).

74. *Xuexi Lu Xun—Shenru pi xiu* (Beijing, 1972), vol. 1, p. 7.

75. LXYJ, 8.378.

76. Yi, pp. 21–22.

77. *Lu Xun pi Kong fan Ru wenji* (Beijing, 1974), and Wang Erling, *Jicheng Lu Xun de fan Kong douzheng chuantong* (Huhehaote, 1974).

78. LXYJ, 8.378.

79. Ibid., 2.337.

80. Jin, p. 420.

81. Yi, p.

82. Zhang Mengyang, *Lu Xun zawen yanjiu liushi nian* (Hangzhou, 1986), p. 127.

83. LXYJ, 6.209–210.

84. Zhang Mengyang, p. 131.

85. See *Jinian Lu Xun dansheng yi bai zhounian wenxian ziliao ji* (Beijing, 1983), pp. 249–330.

86. LXYJ, 8.347, 361.

87. Zhang Mengyang, p. 188.

88. Wang Hui, "Lu Xun yanjiu de lishi pipan," in *Wenxue pinglun* (Beijing), No. 6, 1988, p. 4.

89. Ibid., pp. 4–5.

90. Ibid., p. 4.

91. Ibid., p. 5.

92. Ibid., p. 7.

93. Ibid., p. 5. See also [*Lu Xun yulu*], p. 4.

94. See [*Lu Xun yulu*], pp. 4, 2, and Wang Hui, p. 5.

95. Wang Hui, p. 7.

96. Ibid., p. 5.

97. *Deng Xiaoping wenxuan* (Beijing, 1983), pp. 150–151.

98. Wang Hui, p. 4 (my italics).

99. LXYJ, 6.220.

100. LX, 3.12.

101. LXYJ, 6.214, 218.

102. Ibid., 2.338.

103. Ibid., 6.210.

104. Ibid., 2.338.

105. Wang Hui, p. 8.

106. Shan Yanyi, *Lu Xun yu Qu Qiubai* (Tianjin, 1986), pp. 39, 41, 39. Qu Qiubai was one of Lu Xun's two closest Communist friends, the other being Feng Xuefeng (see below, note 109). An unlikely Communist leader, having been interested primarily in poetry and Buddhism before becoming a Russian language expert, he was pulled into the Party after serving as a translator for Chinese Communists in the Soviet Union, where he had gone for a two year stint as a reporter. Pulled also into intra-Party politics, after the disaster of 1927 and the dismissal of Chen Duxiu, he even found himself head of the Party, for one unfortunate year. Eased out of that

office, relatively gracefully, he "retired" to Shanghai, where he carried out "underground" literary work from 1931 to 1933. Introduced to Lu Xun by Feng Xuefeng, he and his wife thrice sought refuge for several weeks at a time with Lu Xun, when Guomindang agents seemed particularly threatening. According to their wives, Lu Xun and Qu Qiubai became fast friends at their very first meeting. But they only had a year to work together and enjoy each other's company. In 1933 Qu Qiubai went to Ruijin to work for the Party in its "Jiangxi Soviet." Left behind because of illness, when the communists set off on their Long March, he was captured in 1935 by the Guomindang—and shot. See Donald W. Klein and Anne B. Clark, ed., *Biographic Dictionary of Chinese Communism 1921–1965* (Cambridge, 1971), vol. 1, pp. 239–244, and Shan, pp. 68–69.

107. *Lu Xun nianpu*, Lu Xun Bowuguan Lu xun yanjiushi, ed. (Beijing, 1984), 3.65.

108. Yi, p. 7.

109. Rou Shi, who introduced Feng Xuefeng to Lu Xun, seemingly ought to have a prior claim to being Lu Xun's "first close Communist friend," but I *think* Feng Xuefeng joined the Party first. See Goldman, p. xi and Lin, Liu, p. 243. Rou Shi was a young writer (who had audited one of Lu Xun's courses in Beijing) who colaborated with Lu Xun on several projects from 1928 until 1931, when Rou Shi was arrested and secretly executed, with twenty-three others, by the Guomindang.

110. Yi, p. 8.

111. Feng Xuefeng, *Lunwen ji* (Beijing, 1981), I, 3, 6.

112. *Lu Xun shengping shiliao huibian*, Xue Suizhi, chief ed. (Tianjin, 1986), fifth collection, vol. 2, p. 652.

113. See above, n. 109.

114. Yi, p. 10.

115. Feng, I, 33.

116. Yi, p. 10.

117. LX, 4.4–6.

118. Shan, p. 112.

119. Ibid., pp. 76, 91.

120. *Lu Xun shengping shiliao*, pp. 66–67.

121. Feng I, 131, 134 (my italics).

122. [*Lu Xun yulu*], pp. 3, 2.

123. *Mao xuan*, 2.658.

124. Shan, p. 87.

125. LXYJ, 3.255.

126. Yi, p. 10.

127. See the extremely useful article by Wang Junji and Gao Mingluan, "Jin nian lai de Lu Xun sixiang yanjiu gongzuo" (Research work in recent years on Lu Xun's thought), in LXYJ, 3.254–279.

128. Ibid., 3.276.

129. LX, 3.457.

130. See above, p. 71, n. 21.

131. LX, 4.202–203.

132. Ibid., 4.569.

133. ibid., 13.163.

134. Zhang Zhuo, p. 105.

135. Ibid., p. 42.

136. LX, 5.246, 248.

137. Ibid., 6.283.

138. O, p. 489.

139. *Mao yulu*, pp. 174–175.

140. Zhang Zhuo, p. 73.

141. Ibid., p. 101, or Yi, p. 73.

142. I assume the *locus classicus* of this notion is somewhere in the works of Marx. Mao Zedong said, "The socialist system will eventually supercede the capitalist system. This is an objective law that does not turn on the will of people themselves" (*Mao yulu*, p. 22). On June 9, 1989, Deng Xiaoping said of the "turmoil" of May and June of that year: "This storm had to come sooner or later. It was determined by the greater international climate and the lesser climate in China itself. It was bound to come. It did not turn on people's will" (*Renmin ribao haiwai ban*, June 28, 1989, p. 1). His response, he implied, was equally natural.

143. LXYJ, 2.169.

144. Ibid., 2.167.

145. LX, 4.191.

146. Lu Xun told Feng Xuefeng that in 1907, under the influence of "evolutionary thought," he "advocated resistence and national revolution." He "told people to watch out for natural selection and advocated struggling to survive" (see *Lu Xun shengping shiliao*, p. 67). He also said in a famous passage in 1924: "Alas, it is too hard to change China. It almost takes blood even to move a table or put in a new stove" (see LX, 1.164).

147. See LX, 6.151. Lu Xun said of Xiao D: "He is called [lit.] 'Little Same.' When he grows up, he will be just like Ah Q."

148. LX, 4.203.

149. Ibid., 4.191.

150. Ibid., 6.626.

151. LXYJ, 3.274–275.

152. See LXYJ, 8.349. See also Gan Jingcun's excellent article (note his splendid Darwinian name, "Struggle for Existence Gan," or, if one takes literally his surname, admittedly not often a good idea, "Willing to Struggle to Survive"), "'Ci hou zui yaojinde shi gaige guominxing'" (Hereafter the most important thing is to reform the national character), in LXYJ, 3.134.

153. See, for example, two articles from the spring of 1932, "Lin Keduo *Sulian wenjian lu xu*" (A preface to Lin Ke-duo's *A Record of What I Heard and Saw in the Soviet Union*) and "Women bu zai shou pian le" (We shall not be fooled again), LX 4.424–431.

154. See, for example, Lu Xun's preface to his translation of four essays by George Valentinovitch Plekhanov. LX, 4.253–265.

155. See his lament in a letter to Yang Jiyun in 1934: "The lap dog types are not to be feared. The most fearsome are indeed one's 'comrades-in-arms,' who say they are for you, but in their hearts are against you, for there is no way to defend oneself against them." LX, 12.606.

156. *Renmin ribao*, October 20, 1980, p. 8.

157. Wang Hui, p. 13.

158. Ibid., p. 15.

159. See again LX, 1.419, 422, 512–521.

160. See LX, 6.626, and above, pp. 118–119, xiv, 160.

161. Wang Hui, p. 15.

162. Ibid., pp. 14, 6.

163. Ibid., pp. 14–15.

164. See above, pp. 123–127.

165. Wang Hui, pp. 6–7.

166. Joseph R. Levenson, *Confucian China and Its Modern Fate: A Trilogy* (Berkeley and Los Angeles, 1968), III, 82.

167. LX, 5.586.

168. *Mao xuan*, 3.801.

169. Wang Hui, p. 7.

170. LXYJ, 2.338. The Tiananmen Incident of April 5, 1976, followed several days of demonstations, around China's "Memorial Day," in memory of Premier Zhou Enlai, who had died on January 8, 1976, in support of Deng Xiaoping, whom Zhou Enlai had tried to set up as his successor, and against the Gang of Four, who were clearly bent on running things. On the night of April 4, the Gang of Four sent security police to Tiananmen Square to clear out the thousands of wreathes, banners, and poems left there by the demonstrators. On the night of April 5, police returned to drive out the demonstrators who remained, after a day of angry protests against the destruction of their wreathes. Witnesses have reported beatings and bloodshed and arrests, but to my knowledge there are no confirmed fatalities. The Tiananmen Incident of June 4, 1989, followed weeks of student demonstrations against corruption and for democracy, demonstrations finally crushed on the night of June 4, by soldiers of the People's Liberation Army. This time there was appalling bloodshed. Estimates of the dead range from 400 to 3000.

171. LX, 5.382.

172. Ibid., 3.407.

173. Ibid., 13.155.

174. Ibid., 1.350.

175. Ibid., 6.529–530.

176. Ibid., 13.155.

177. Ibid.

178. Ibid., 7.116.

179. A common classical idiom: "*yang hu yi huan.*"

180. Cf. Liang Qichao's famous formula: "*bian yi bian, bu bian yi bian*" (If we change, we shall change, if we do not change we shall change). Liang Qichao, *Yin bing shi*, I, 1.8.

181. LX, 2.198.

182. See above, Chapter 8, n. 117.

183. Huxley, *Man's Place*, p. 132.

184. O, p. 488.

185. Once again LX, 1.422, the first line of "the madman's" diary.

186. Huxley, *Man's Place*, p. 132.

Glossary

Ah Q zheng zhuan 阿Q 正传
ai guo 爱国
Aiguo xueshe 爱国学社
Ba Ren 巴人
bai sheng 白眚
banzai 万岁
baoluan zhi ren 暴乱之人
baoluan zhi tu 暴乱之徒
ben xing 本性
bian yi bian, bu bian yi bian 变亦变，不变亦变
biange 变革
Bo Yi 伯夷
bu zheng ba 不争霸
bushido 武士道
Cai Yuanpei 蔡元培
Cao Cao 曹操
cha renlei zhi bu qi 察人类之不齐
Chan 禅
Chen Boda 陈伯达
Chen Duxiu 陈独秀
Chen Heng 陈恒
Chen Yong 陈涌
Cheng Ke 程克
chi ren 吃人
chizi zhi xin 赤子之心
Chou Chih-p'ing (Zhou Zhiping) 周质平
"'Ci hou zui yaojinde shi gaige guomin xing'" 此后最要紧的是改革国民性
cibei 慈悲
da tong 大同
dang tong fa yi 党同伐异
danwei 单位
dao 道
de zhi zei 德之贼
Demokelaxi Xiansheng 德谟克拉西先生
Deng Xiaoping 邓小平
diguozhuyizhe 帝国主义者
Ding Ling 丁玲

Duan Qirui 段祺瑞
dui diren de cibei jiushi dui renmin de canren 对敌人的慈悲就是对人民的残忍
Er xin ji 二心集
Er yi ji 而已集
fa cibei 发慈悲
Fan Chi wen ren. Zi yue 'Ai ren.' 樊迟问仁 。子曰爱人
fan zhu ji 反诸己
fei yue 飞跃
Fen 坟
"Fengci wenxue yu shehui gaige" 讽刺文学与社会改革
Gan Jingcun 甘竟存
Gao Mingluan 高鸣鸾
Gao Zi 告子
geming 革命
geming bushi qingke chifan 革命不是请客吃饭
"Geming yu zhishi jieji" 革命与知识阶级
Gongsun Long 公孙龙
Gu Lang 顾琅
gu ming si yi 顾名思义
Gu shi xin bian 故事新编
guan bi min fan 官逼民反
"Guan xin shuo" 观心说
Guangfu Hui 光复会
Guangming ribao 光明日报
Guji xuba ji 古籍序跋集
guo wang zhong mie 国亡种灭
guomin xing 国民性
Guomindang 国民党
Han Shaohua 韩少华
Han wenxue shi gangyao 汉文学史纲要
Han Yu 韩愈
he er er yi 合二而一
Henan 河南
Hou Yi 后羿
Hu Feng 胡风
Hu Shi 胡适
Hua bian wenxue 花边文学
Hua gai ji 华盖集
Hua gai ji xubian 华盖集 续编
hua xing 化性
huchou 狐臭
Ji Kang 嵇康
Ji Kang Zi 季康子
Ji wai ji 集外集
Ji wai ji shiyi 集外集拾遗
Ji wai ji shiyi bubian 集外集拾遗补编

jian ai 兼爱

Jiang Jieshi 蒋介石

Jiang Qing 江青

jiang shan yi gai, ben xing nan yi 江山易改，本性难移

"Jiaru ta hai huozhe" 假如他还活着

jie gu feng jin 借古讽今

jimo er wu sheng 寂漠而无声

"Jin nian lai de Lu Xun si xiang yanjiu gongzuo" 今年来的鲁迅思想研究工作

jingshen jie zhi zhanshi 精神界之战士

Jingsong 景宋

jinhua lun 进化论

junzi 君子

Kang Sheng 康生

ke ji 克己

ke ji fu li wei ren 克己复礼为仁

"Kexue shi jiao pian" 科学史教篇

Kong Rong 孔融

Kong Yiji 孔乙己

"Kuangren riji" 狂人日记

laozi yingxiong er haohan, laozi fandong er hundan 老子英雄儿好汉，老子反动儿浑蛋

Lei Feng 雷锋

Lei Jun 雷军

Li Helin 李何林

li ren 立人

Li Shizeng 李石曾

liang bian 量变

Liang di shu 两地书

Liang Shiqiu 梁实秋

Liang Xiao 梁校

lie deng 躐等

Lin Biao 林彪

"Lin Keduo *Sulian wenjian lu* xu" 林克多《苏联闻见录》序

linggan 灵感

Liu Hezhen 刘和珍

Liu Kunyi 刘坤一

Liu Shaoqi 刘少奇

Liu Shipei 刘师培

"Lu Xun yanjiu de lishi pipan" 鲁迅研究的历史批判

"Lun 'feie polai' yinggai huan xing" 论《费厄泼赖》应该缓行

"Lun shang wu" 论尚武

luo chong 倮虫

Luo Feng 罗烽

Ma Junwu 马君武

"Ma sha yu peng sha" 骂杀与捧杀

mamu buren 麻木不仁

Masuda Sho 增田涉

Min bao 民报
ming bu zheng ze yan bu shun 名不正则言不顺
"*Moluo shi li shuo*" 摩罗诗力说
Nahan 呐喊
Nan qiang bei diao ji 南腔北调集
neibu 内部
nimen keyi gai le 你们可以改了
Ouyang Shan 欧阳山
pai 派
Panghuang 彷徨
pimao 皮毛
"*Po e sheng lun*" 破恶声论
Pu Liangpei 濮良沛
qi yi cong wen 弃医从文
Qian Dajun 钱大钧
qie er bu she 锲而不舍
Qie jie ting zawen 且介亭杂文
Qie jie ting zawen er ji 且介亭杂文二集
Qie jie ting zawen mobian 且介亭杂文末编
Qin Shi Huang 秦始皇
"*Qingqidui*" 轻骑队
Qiu Jin 秋瑾
qu gao he gua 曲高和寡
Qu Qiubai 瞿秋白
quchong 蛆虫
Re feng 热风
ren (human being) 人
ren (benevolence) 仁
ren ren 仁人
ren wei 人为
ren ye zhe ren ye 仁也者人也
ren zhe ren ye 仁者人也
"*Ren zhi lishi*" 人之历史
rendao 人道
rendaozhuyizhe 人道主义者
renrenmen 仁人们
renxing 人性
renxing de zhanzheng 韧性的战争
renyi daode 仁义道德
Rou Shi 柔石
Ruan Ji 阮藉
Ruo 若
ruo rou qiang shi 弱肉强食
Saiyinsi Xiansheng 赛恩斯先生
San xian ji 三闲集
seppuku 切腹

shan he 山河
Shen Diemin 沈瓞民
shen shi 神矢
shenghua 圣化
shengjue 圣觉
shengren 圣人
shou xing zhi ai guo 兽性之爱国
shu 恕
Shu Qi 叔齐
Shui hu zhuan 水浒传
Shun 舜
"Shuo ri" 说铟
"Sibada zhi hun" 斯巴达之魂
Sima Qian 司马迁
Sima Yi 司马懿
Song Qingling 宋庆龄
Song Yu 宋玉
Su bao 苏报
Sun Yatsen (Sun Yixian) 孙逸仙
Sun Zhudan 孙竹丹
Tian Han 田汉
tianyan lun 天演论
Tong Meng Hui 同盟会
tou bi cong rong 投笔从戎
wan wu zhi ling 万物之灵
Wang Chong 王充
Wang Furen 王富仁
Wang Hongwen 王洪文
Wang Jinfa 王金发
Wang Junji 王骏骥
Wang Ruowang 王若望
Wang Shiwei 王实味
Wang Weixin 王维新
Wei ziyou shu 伪自由书
"Wenhua pian zhi lun" 文化偏至论
wenyan 文言
Wo suo renshi de Lu Xun 我所认识的鲁迅
"Women bu zai shou pian le" 我们不再受骗了
wuhu 呜呼
"Wu sheng de Zhongguo" 无声的中国
Wu Wang 武王
Wu Zetian 武则天
Wu Zhihui 吴稚晖
wu zhong sheng you 无中生有
xian zhi xian jue 先知先觉
xianshengmen 先生们

Xiang Yu 项羽
Xianglin Sao 祥林嫂
xiangyuan 乡 愿
xiao 孝
Xiao Hong 萧红
Xiao Jun 萧军
Xin Qingnian 新青年
Xin sheng 新生
xin sheng 新生
Xin shiji 新世纪
Xu Maoyong 徐懋庸
Xu Xilin 徐锡麟
Yang Hansheng 阳翰笙
yang hu yi huan 养虎遗患
Yang Jiyun 杨霁云
Yang Quan 杨铨
Yang Zhihua 杨之华
Yao 尧
Yao Wenfu 姚文甫
Yao Wenyuan 姚文元
Ye cao 野草
"'Yi fen wei er' yu 'he er er yi'" 一分为二与合二而一
yi qiu zhi he 一丘之貉
yi xin qiu xin 以心求心
yinyang 阴阳
Yiwen xuba ji 译文序跋集
Yu 禹
Yuan Shikai 袁世凯
zao fan you li 造反有理
zawen 杂文
Zhang Binglin 章炳麟
Zhang Chunqiao 张春桥
Zhang Shizhao 张士钊
Zhang Yu'an 张玉安
Zhang Zhidong 张之洞
"Zhanshi he cangying" 战士和苍蝇
Zhao hua xi shi 朝花夕拾
zhen de ren 真的人
zhengren junzi 正人君子
zhengyide 正义的
zhengyigan 正义感
zhengzhi gua shuai 政治挂帅
zhi bian 质变
zhijue 直觉
zhong 忠
"Zhongguo dizhi luelun" 中国地质略论

Zhongguo Jiaoyu Hui   中国教育会
Zhongguo kuangchan quan tu 中国矿产全图
Zhongguo kuangchan zhi 中国矿产志
Zhongguo Minquan Baozhang Tongmeng   中国民权保障同盟
Zhongguo renmin dao liao zui weixian di shihou 中国人民到了最危险的时候
Zhongguo xiaoshuo shilue 中国小说史略
"Zhongguo zhi hun an zai hu" 中国之魂安在乎
Zhou Enlai   周恩来
Zhou Gong   周公
Zhou Jianren   周建人
Zhou Shuren (Lu Xun) 周树人
Zhou Yang   周扬
Zhou Zuoren   周作人
Zhu Xi   朱熹
Zhu Zheng   朱正
Zhun feng yue tan 准风月谈
zi qi qi ren 自欺欺人
zi wei zheng, yan yong sha 子为政，焉用杀
Zou Rong   邹容
zun gexing er zhang jingshen 尊个性而张精神

Bibliography

Aeschyli Septem Quae Supersunt Tragoedias. Denys Page, ed. Oxford, Oxford University Press, 1972.

Alistair Cooke's America. New York, Alfred A. Knopf, 1973.

Appleton, Philip, ed. *Darwin.* New York, Norton, 1979.

Ayers, William. *Chang Chih-tung and Educational Reform in China.* Cambridge, Harvard University Press. 1971.

Barlow, Nora, ed. *The Autobiography of Charles Darwin.* New York, Norton & Co., 1969.

Camus, Albert. *The Myth of Sisyphus.* Justin O'Brien, trans. New York, Random House, 1955.

Cheng Ma 程麻. "Lu Xun liu Ri shiqi de qimengzhuyi sixiang jiegou" 鲁迅留日时期的启蒙主义思想结构 (The structure of Lu Xun's enlightenment thought in the period of his study in Japan), in *Lu Xun yanjiu*, 8 (see below).

Chow Tse-tsung 周策纵 . *The May Fourth Movement.* Cambridge, Harvard University Press, 1968.

Ci hai 辞海 (A sea of words). Shanghai, 1979.

Dai Qing 戴晴. *Wang Shiwei and "Wild Lilies."* New York, M.E. Sharpe, 1994.

Darwin, Charles. *On the Origin of Species.* Facsimile of 1st ed. Cambridge, Harvard University Press, 1966.

———. *The Descent of Man.* New York & London, Merrill & Baker, n.d.

———. *The Expression of the Emotions in Man and Animals.* Chicago, The University of Chicago Press, 1965.

———. *The Voyage of the Beagle.* New York, Doubleday Anchor, 1962

Darwin, Francis. *The Life and Letters of Charles Darwin.* New York, D. Appleton and Co., 1889.

Dawkins, Richard. *The Selfish Gene.* New York, Oxford University Press, 1978.

De Tocqueville, Alexis. *Democracy in America.* 2 vols. New York, Vintage Books, 1945.

Deng Xiaoping wenxuan 邓小平文选 (Selected works of Deng Xiaoping). Beijing, 1983.

Dostoyevsky, Fyodor. *Crime and Punishment*. London (?), Penguin Books, 1988.

Dr. Seuss. *Horton Hatches the Egg*. New York, Random House, 1940.

Eastman, Lloyd E. *The Abortive Revolution*. Cambridge, Harvard University Press, 1975.

Eiseley, Loren. *Darwin's Century*. New York, Doubleday Anchor, 1958.

Feng Xuefeng 冯雪峰 . *Lunwen ji* 论文集 (Collected articles). 3 vols. Beijing, 1981.

Flanders, Michael and Donald Swan. *At the Drop of a Hat*. London, Angel Records S.35797, n.d.

———. *The Bestiary of Flanders and Swan*. London, Angel Records, S.36112, n.d.

Fokkema, D. W. *Literary Doctrine in China and Soviet Influence 1956–1960*. The Hague, Mounton & Co., 1965.

"Freedom on the Wallaby." Henry Lawson, words, on Gordon Bok, *Bay of Fundy*, Sharon Connecticut, Folk-Legacy Records, 1975.

Freud, Sigmund. *Civilization and Its Discontents*. New York, Norton, 1962.

Goldman, Merle. *Literary Dissent in Communist China*. Cambridge, Harvard University Press, 1967.

Gould, Stephen Jay. *Ever Since Darwin*. New York, Norton, 1979.

———. *Hens' Teeth and Horses' Toes*. New York, Norton, 1983.

———. *Ontogeny and Phylogeny*. Cambridge, Belknap Press of Harvard University Press, 1977.

———. *The Flamingo's Smile*. New York, Norton, 1985.

———. *The Panda's Thumb*. New York, Norton, 1982.

Gray, Asa. *Darwiniana*. Cambridge, The Belknap Press of Harvard University Press, 1963.

Haeckel, Ernst. *Monism*. J. Gilchrist, tr. London, Adam and Charles Black, 1895.

———. *The Evolution of Man* (tr. of *Anthropogenie*). 2 vols. New York, D. Appleton, 1897.

———. *The History of Creation* (tr. of *Naturliche Schopfungsgeschichte*, 1868), "A young lady," tr. Tr. revised by E. Ray Lankester. 2 vols. New York, D. Appleton, 1897.

———. *The Riddle of the Universe* (Tr. of *Die Weltsatsel*). Joseph McCabe, tr. New York, Harper & Brothers, 1900.

Handel, G. F. *The Messiah*. New York, Schirmer, 1912.

Herrstein, Richard. "I.Q.," *The Atlantic Monthly* 228 (1971), pp. 43–64. (Quoted in part in *Darwin*, Philip Appleman, ed., pp. 490–499. See above).

Hersey, John. *The Call*. New York, Alfred A, Knopf, 1985.

Himmelfarb, Gertrude. *Marriage and Morals among the Victorians*. New York, Alfred A. Knopf, 1986.

————. *Victorian Minds*. New York, Harper & Row, 1970.

Hofstadter, Richard. *Social Darwinism in American Thought*, rev. ed. Boston, Beacon Press, 1955.

Hu Sheng 胡绳 . *Zao xia lun cong* 枣下论丛 (Studies under the date tree). Beijing, 1978.

Hugo, Victor. *Oeuvres Politiques complètes—Oeuvres Diverses*. Jean-Jacques Pauvert, ed. Paris, 1964.

Huiyi Lu Xun ziliao jilu 回忆鲁迅资料辑录 (Topically organized recollections of Lu Xun), Shanghai, 1980.

Huxley, Thomas H. *Evolution and Ethics*. London, MacMillan and Co., 1895.

————. *Man's Place in Nature*. Ann Arbor, University of Michigan Press, 1959.

————. *Science and Christian Tradition*. London, MacMillan and Co., 1897.

Irvine, William. *Apes, Angles, and Victorians*. New York, McGraw-Hill, 1955.

Ji Kang ji 嵇康集 (Collected works of Ji Kang). Lu Xun, ed. Hong Kong, 1967

Jin Hongda 金宏达 . *Lu Xun wenhua sixiang tansuo* 鲁迅文化思想探索 (An exploration of Lu Xun's cultural thought). Beijing, 1986.

Jinian Lu Xun dansheng yi bai zhounian wenxian ziliao ji 纪念鲁迅诞生一百周年文献资料集 (A collection of documentary materials commemorating the one hundredth anniversary of Lu Xun's birth). Beijing, 1983.

Kelly, Walt. *Pogo*. New York, Simon and Schuster, 1951.

————. *Pogo: We Have Met the Enemy and He Is Us*. New York, Simon and Schuster, 1972.

Klein, Donald W. and Anne B. Clark, eds. *Biographic Dictionary of Chinese Communism, 1921–1965*. 2 vols. Cambridge, Harvard University Press, 1971.

Kristof, Nicholas D. and Sheryl WuDunn. *China Awakes*. New York, Random House, 1994.

Kropotkin, Peter A. *Memoirs of a Revolutionist*. New York, Grove Press, 1970.

————. *Mutual Aid*. Popular ed. London, William Heinemann, 1919.

Lao Zi dao de jing 老子道德经, in *Wang Bi ji jiaoshi* 王弼集较释 (Wang Bi's collected works, annotated). Beijing, 1980.

Levinson, Joseph R. *Confucian China and Its Modern Fate: A Trilogy*. Berkeley and Los Angeles, University of California Press, 1968.

Leys, Simon (Pierre Ryckmans). *Chinese Shadows*. New York, The Viking Press, 1977.

Li Zehou 李泽厚. *Zhongguo jindai sixiang shi lun* 中国近代思想史论 (Discussions of Modern Chinese intellectual history). Beijing, 1979.

Liang Qichao 梁启超. *Xin min shuo* 新民说 (Towards a new people), Taipei, 1959.

———. *Yin bing shi wenji* 饮冰室文集 (Collected writings from the ice-drinker's studio). 16 vol. Taipei, 1960.

———. *Ziyou shu* 自由书 (On liberty). Taipei, 1960.

Lin Fei 林非, Liu Zaifu 刘再复. *Lu Xun zhuan* 鲁迅传 (a biography of Lu Xun). Beijing, 1981.

Linnaeus, Carolus. *Systema Naturae*. A photographic facsimile of the first volume of the tenth edition (1758). London, printed by order of the Trustees, British Museum (Natural History), 1956.

Liu Zaifu 刘再复, Jin Qiupeng 金求鹏, Wang Zichun 汪子春. *Lu Xun he ziran kexue* 鲁迅和自然科学 (Lu Xun and natural science). Beijing, 1979.

Lorenz, Konrad. *On Aggression*. New York, Harcourt, Brace & World, 1963.

Lu Xun 鲁迅. *Fen* 坟 (Tomb). Hong Kong, 1964.

Lu Xun nianpu 鲁迅年谱 (A chronology of Lu Xun's life) Fudan Daxue, Shanghai Shida, Shanghai Shiyuan *Lu Xun nianpu* bianxie zu 复旦大学, 上海师大, 上海师院. 鲁迅年普遍写组 (The Fudan University, Shanghai Normal University and Shanghai Teachers' College Lu Xun nianpu editorial group), ed. Hefei, 1979.

Lu Xun nianpu 鲁迅年谱 (A chronology of Lu Xun's life). Vols. 3 and 4. Lu Xun Bowuguan Lu Xun yanjiu shi 鲁迅博物馆鲁迅研究室 (The Lu Xun research office of the Lu Xun Museum), ed. Beijing, 1984.

Lu Xun pi Kong fan ru wenji 鲁迅批孔反儒文集 (Selected Lu Xun texts criticizing Confucius and opposing Confucianists). Beijing, 1974.

Lu Xun quan ji 鲁迅全集 (The complete works of Lu Xun). 16 vols. Beijing, 1981.

Lu Xun shengping shiliao huibian 鲁迅生平史料汇编 (A compilation of historical materials of Lu Xun's life). Fifth collection. 2 vols. Xue Suizhi 薛绥之, chief ed. Tianjin, 1986.

Lu Xun shi ge zhu 鲁迅诗歌注(Annotated poems and songs of Lu Xun). Zhou Zhenfu 周振甫, ed. Hangzhou, 1980.

Lu Xun shi jian xuanji 鲁迅诗笺选集 (Selected annotated poems of Lu Xun). Wenxue yanjiu she 文学研究社 ed. Hong Kong, 1967.

Lu Xun yanjiu 鲁迅研究 (Lu Xun research), vol. 1. Lu Xun yanjiu editorial devision of the Lu Xun Research Association, ed. Shanghai, 1980.

———. Vol. 2. Chinese Lu Xun Research Society's Lu Xun yanjiu Editorial Board, ed. Beijing, 1981.

———. Vol. 3. Beijing, 1981.

———. Vol. 6. Beijing, 1982

———. Vol. 8. Beijing, 1983.

Lu Xun yanjiu ziliao 鲁迅研究资料 (Lu Xun research materials), vol. 7. Beijing Lu Xun Museum Lu Xun Research Office, ed. Tianjin, 1980.

———. Vol. 17. Tianjin, 1986.

Lu Xun yanlun xuanji 鲁迅言论选集 (Selected pronouncements of Lu Xun), vol. 3. Beijing, 1976.

[*Lu Xun yulu*] 鲁迅语录 (Quotations from Lu Xun). An untitled selection of quotations from Lu Xun, published in same format as *Mao zhuxi yulu* (See below). No ed. No pub. N.d. Published in China after 1966, probably late in the Cultural Revolution.

Lu Xun zawen xuanjiang 鲁迅杂文选讲 (Selected essays of Lu Xun). n.p., 1973.

Lu Xun zhi Xu Guangping shujian 鲁迅致许广平书简 (Lu Xun's letters to Xu Guangping). Shijiazhuang, 1979.

Lun heng zhushi 论衡注释 (Lun Heng annotated). Peking University History Department *Lun Heng* annotation committee, ed., 4 vols. Beijing, 1979.

Lunyu yizhu 论语译注 (The Analects with notes and vernacular translation). Yang Bojun 杨伯峻, ed. Beijing, 1965.

Lutz, Jessie G. *Christian Missions in China—Evangelists of What?* Boston, D.C. Heath and Co., 1965.

Lyell, William A., Jr. *Lu Hsun's Vision of Reality*. Berkeley, University of California, 1976.

Ma Tiji 马蹄疾. *Lu Xun jiangyan kao* 鲁迅讲演考 (Lu Xun's speeches). Haerbin, 1981.

Mao Zedong lun wenxue he yishu 毛泽东论文学和艺术 (Mao Zedong discusses art and literature). Beijing, 1964.

Mao Zedong xuanji 毛泽东选集 (Selected works of Mao Zedong). 4 vols. Beijing, 1969.

———. Vol. 5. Shanghai, 1977.

Mao zhuxi yulu 毛主席语录 (Quotations from Chairman Mao). Beijing, 1967.

Meng Zi yi zhu 孟子译注 (Mencius, with vernacular translation and notes). *Meng Zi yi zhu* team of the Lanzhou University Chinese Department, ed. Hong Kong, n.d.

Mill, John Stuart. *On Liberty*. Aldurey Castell, ed. New York, Appleton-Century-Crofts, 1947.

Milton, David and Nancy Milton, Franz Schurman, eds. *People's China*. New York, Random House, 1974.

Min Kangsheng 闵抗生 . "'Guoke' yu *Also Sprach Zarathustra*" 《过客》与 *Also Sprach Zarathustra* ("The Traveler" and Also Sprach Zarathustra). Paper delivered at the Lu Xun and Chinese and Foreign Culture Conference, Beijing, October 19–25, 1986.

Mo Zi jian gu 墨子间诂 (An annotated *Mo Zi*). Sun Yirang, 孙诒让 ed. 2 vols. Taipei, 1970.

Morris, Desmond. *The Naked Ape*. New York, Dell, 1984.

Morris, Richard. *Evolution and Human Nature*. New York, Avon Books, 1984.

Ni Moyan 倪墨炎 . *Lu Xun geming huodong kaoshu* 鲁迅革命活动考述 (An investigative account of Lu Xun's revolutionary activities).

Nietzsche, Friedrich. *Thus Spoke Zarathustra*. R. J. Hollingdale, tr. Baltimore, Penguin Books, 1968.

Pusey, James Reeve. *China and Charles Darwin*. Cambridge, Council on East Asian Studies, Harvard University, 1983.

———. *Wu Han: Attacking the Present Through the Past*. Cambridge, East Asian Research Center, Harvard University, 1969.

Pye, Lucian. *China: An Introduction*. Boston, Little Brown, 1972.

Qing yi bao 清议报 (Honest criticsm). Liang Qichao, Feng Jingru 冯镜如 et al., eds. Photolithograph. 12 vols. Taipei, 1967.

Rankin, Mary Backus. *Early Chinese Revolutionaries*. Cambridge, Harvard University Press, 1971.

Renmin ribao 人民日报 (People's Daily). Beijing, 1980.

Renmin ribao haiwai ban 人民日报海外版 (People's Daily overseas edition). Beijing, 1989.

Rubin, Kyna. "Keeper of the Flame: Wang Ruowang as Moral Critic of the State," in Merle Goldman, ed. with Timothy Cheek and Carol Lee Hamrin, *China's Intellectuals and the State: In Search of a New Relationship*. Cambridge, Council on East Asian Studies/ Harvard University, 1987.

Sartre, Jean-Paul. *Existentialism and Human Emotions*. New York, 1957.

Schram, Stuart, ed. *Chairman Mao Talks to the People*. New York, Random House, 1974.

Schwartz, Benjamin. *In Search of Wealth and Power: Yen Fu and the West*. Cambridge, Harvard University Press, 1964.

Shan Yanyi 单演义 . *Lu Xun yu Qu Qiubai* 鲁迅与瞿秋白 (Lu Xun and Qu Qiubai). Tianjin, 1986.

Shi ji hui zhu kaozheng (Shiki kaichu kosho) 史记会注考证 (Shi ji with selected notes from various commentaries, critically assessed). Takigawa Kametaro 泷川龟太郎 , ed. N.p., n.d.

Skinner, B. F. *Walden Two*. New York, Macmillan, 1976.

Spencer, Herbert. *Social Statics*. New York, D. Appleton and Company, 1896.

———. *Universal Progress*. New York, Appleton and Company, 1864.

Su Shuangbi 苏双碧 and Wang Hongzhi 王宏志 . *Wu Han zhuan* 吴晗传 (A biography of Wu Han). Beijing, 1984.

The Chinese-English Dictionary. English Department, Beijing Foreign Language Institute, ed. Hong Kong, Commercial Press, 1979.

The Confessions of St. Augustine. James Marshall Campbell and Martin R. P. McGuire, eds. Englewood Cliffs, Prentice Hall, 1959.

The Greek New Testament. Kurt Aland, et al. eds. New York, United Bible Societies, 1968.

The Songs of Michael Flanders and Donald Swan. New York, St. Martin's Press, 1977.

Wang Erling 王尔龄 . *Jicheng Lu Xun de fan Kong douzheng chuantong* 继承鲁迅的反孔斗争传统 (Carry on the tradition of Lu Xun's struggle against Confucius). Huhehaote, 1974.

Wang Hui 汪晖 . "Lu Xun yanjiu de lishi pipan" 鲁迅研究的历史批判 (An historical criticsm of Lu Xun research), *Wenxue pinglun* 文学评论 (Literary review), No. 6. Beijing, 1988.

Wang Shiqing 王士菁 . *Lu Xun chuangzao daolu chu tan* 鲁迅创造道路初探 (An initial exploration of Lu Xun's creative way). Beijing, 1981.

———. *Lu Xun zao qi wu pian lunwen zhuyi* 鲁迅早期五篇论文注译 (Five of Lu Xun's early essays, with notes and translation). Tianjin, 1981.

———. *Lu Xun zhuan* 鲁迅传 (A biography of Lu Xun). Beijing, 1981.

Weiner, Jonathan. *The Beak of the Finch*. New York, Alfred A. Knopf, 1994.

Wenxue yundong shiliao xuan 文学运动史料选 (Selected historical materials from the literature movement), vol. 4. Shanghai, 1979.

Wilson, Edward O. *On Human Nature*. New York, Bantam, 1979.

———. *Sociobiology: The New Synthesis*. Cambridge, The Belknap Press of Harvard University Press, 1975.

Wu Han 吴晗. *Chuntian ji* 春天集 (Spring collection). Beijing, 1961.

———. *Deng xia ji* 灯下集 (Lamplight collection). Beijing, 1957.

Xin min congbao 新民丛报 (A new people). Feng Zishan 冯紫珊, ed. Photolithograph. 17 vols. Taipei, 1966.

Xu Guangping yi Lu Xun 许广平忆鲁迅 (Xu Guangping's reminiscences of Lu Xun). Ma Tiji, ed. Guangzhou (?), 1979.

Xu Shoushang 许寿裳. *Wang you Lu Xun yinxiang ji* 亡友鲁迅印象记 (Recorded impressions of my late friend Lu Xun). Beijing, 1981.

Xuexi Lu Xun—Geming dao di 学习鲁迅 革命到底 (Study Lu Xun—Revolt to the end). Shanghai, 1973.

Xuexi Lu Xun—Shenru pi xiu 学习鲁迅　深入批修 (Study Lu Xun—Deeply criticize revisionism), vol. 1. Beijing, 1972.

Xun Zi jian shi 荀子柬释 (*Xun Zi*, with selected explanations). Liang Qixiong 梁启雄, ed. Taipei, 1969.

Yan Fu 严复. *Tianyan lun* 天演论 (The theory of natural evolution). Taipei, 1967. A translation of Thomas Henry Huxley (Hexuli 赫胥黎), *Evolution and Ethics*.

Yi Zhuxian 易竹贤. *Lu Xun sixiang yanjiu* 鲁迅思想研究 (Research on Lu Xun's thought). Wuhan, 1984.

Yue Daiyun and Carolyn Wakeman. *To the Storm*. Berkeley, University of California Press, 1985.

Zhang Mengyang 张梦阳. *Lu Xun zawen yanjiu liushi nian* 鲁迅杂文研究六十年 (Sixty years of research on Lu Xun's random essays). Hangzhou, 1986.

Zhang Zhaoyi 张钊贻. "Jinhua lun yu chaoren de maodun" 进化论与超人的矛盾 (The contradiction between the theory of evolution and the superman), in *Lu Xun yanjiu ziliao*, vol. 17 (see above).

Zhang Zhuo 张琢. *Lu Xun zhexue sixiang yanjiu* 鲁迅哲学思想研究 (A study of Lu Xun's philosophical thought). Hubei Province, 1981.

Zhao Ruihong 赵瑞蕻. *Lu Xun "Moluo shi li shuo" zhushi, jinyi, jieshuo* 鲁迅《摩罗诗力说》注释, 今译, 解说 (Lu Xun's "The power of Satanic poetry," with notes, a modern translation, and an explanation). Tianjin, 1982.

Zheng Yi 正一. Lu Xun sixiang fazhan lun gao 鲁迅思想发展论稿 (A draft discussion of the development of Lu Xun's thought). Chengdu, 1981.

Zhong yong 中庸 (The doctrine of the mean), in *Si shu xin jie* 四书新解 (A new exposition of the four books). Zhang Shoubai 张守白, ed. Tainan, 1961.

Zhongguo dabaikequanshu, zhexue (The Chinese encyclopedia, philosophy)中国大百科全书, 哲学 . Beijing/Shanghai, 1987.

Zhongguo zhexue shi jiaoxue ziliao xuanji 中国哲学史教学资料选集 (Selected teaching materials in the history of Chinese philosophy). Beijing Daxue zhexue xi Zhongguo zhexue shi jiaoyan shi 北京大学哲学系中国哲学史教研室 (History of Chinese philosophy teaching and research group of the Philosophy Department of Beijing University), ed. 2 vols. Beijing, 1981.

Zhongguo zhexue shi ziliao xuanji, Song, Yuan, Ming zhi bu 中国哲学史资料选集, 宋, 元, 明之部 (Selected source materials in the history of Chinese philosophy: Song, Yuan, and Ming). Zhongguo shehui kexue yuan zhexue yanjiusuo Zhongguo zhexue shi yanjiu shi 中国社会科学院哲学研究所中国哲学史研究室 (The history of Chinese philosophy research office of the Philosophy Research Institute of the Chinese Academy of Social Sciences), ed. Beijing, 1984.

Zhuang Zi ji shi 庄子集释 (Zhuang Zi with collected commentaries). Guo Qingfan 郭庆藩 , ed. 4 vols. Beijing, 1961.

Index